MODERN TRENDS IN TUNNELLING AND BLAST DESIGN

Modern Trends in Tunnelling and Blast Design

By
JOHN JOHANSEN
In association with
C.F. MATHIESEN

A.A. BALKEMA / ROTTERDAM / BROOKFIELD / 2000

Published by
A.A.Balkema, P.O.Box 1675, 3000 BR Rotterdam, Netherlands
Fax: +31.10.4135947; E-mail: balkema@balkema.nl; Internet site: http://www.balkema.nl

A.A.Balkema Publishers, Old Post Road, Brookfield, VT 05036-9704, USA
Fax: 802.276.3837; E-mail: info@ashgate.com

ISBN 90 5809 311 5 hardbound edition
ISBN 90 5809 312 3 student paper edition

Contents

Table of units and conversion factors

length	1 inch (in) = 25,4 mm	1 m = 39,37 in
	1 in = 2,54 cm	1 m = 3,28 ft
		1 m = 1,09 yd
	1 foot (ft) = 304,8 mm	1 cm = 0,39 in
	1 ft = 30,5 cm	
	1 ft = 0,305 m	
	1 yard (yd) = 0,91 m	
area	1 in^2 = 6,45 cm^2	1 m^2 = 1550 in^2
	1 ft^2 = 0,093 m^2	1 cm^2 = 0,15 in^2
	1 ft^2 = 930 cm^2	
	1 yd^2 = 0,836 m^2	
volume	1 yard^3 = 0,765 m^3	1 m^3 = 1,31 yd^3
	1 ft^3 = 0,028 m	1 cm^3 = 0,06 in^3
	1 in^3 = 16,36 cm^3	
time	1 sec (s)	1 second (s)
	1 millisec (ms, MS) = 0,001 s	1 ms = 0,001 s
frequency	cycles per sec (c/s)	cycles (periods) per sec (c/s, p/s)
velocity	1 ft/s = 0,31 m/s	1 m/s = 3,28 ft/s
	1 in/s = 2,54 cm/s	1 cm/s = 0,39 in/s
mass	1 lb = 0,45 kg	1 kg = 2,20 lb
density	1 lb/ft^3 = 16 kg/m^3	1 kg/m^3 = 0,063 lb/ft^3
	= 0,016 g/cm^3	1 kg/dm^3 = 62,5 lb/ft^3
spec. charge	1 lb/yd^3 = 0,59 kg/m^3	1 kg/m^3 = 1,69 lb/yd^3
concentration of charge	1 lb/ft = 1,48 kg/m	1 kg/m = 0,67 lb/ft
acceleration	1 ft/s^2 = 0,31 m/s^2	1 m/s^2 = 3,28 ft/s^2
pressure	1 psi = 0,073 kg/cm^2	1 kg/cm^2 = 14,22 psi

CHAPTER 1

Introduction

Eight to hundred years ago underground construction work was mostly associated with mining activities. In the last decades, however, tunnelling has become a more important activity within the field of underground construction work. Tunnels are driven for road and rail, sewer and water supply, hydro electric power projects, mountain caverns for industrial, recreational and storage purposes and as transport and storage systems for oil and gas. A number of these tunnel sites are underneath or close to inhabited buildings or industrial installations. Such conditions require a good technical judgment and knowledge of tunnel driving together with careful planning, to ensure safe and satisfactory results.

The conventional tunnelling technology is progressing rapidly. The performance capability of the construction equipment, in particular that of the drilling equipment, has increased considerably over the last few years.

With the introduction of more effective and more accurate hydraulic drilling rigs, with higher capacity of the loading and transportation equipment together with modern and more advanced explosives and detonating systems, loading and blasting techniques have increased the speed of progress in tunnelling and reduced construction time and cost. 70-80 m of tunnel advance per week is today quite normal for road and railway tunnels.

CHAPTER 2

General

2.1 THE TUNNEL PORTAL AND INITIAL APPROACH

When the tunnel portal has been fixed, the area must be carefully cleared of soil and loose rock. The portal is then blasted to give an overburden at the entrance of at least half times the tunnel diameter.

The first round is blasted very carefully with minimum borehole depth. In poor quality rock it may even be necessary to start with a smaller pilot tunnel which is subsequently enlarged to the full tunnel dimension (Fig. 2.1).

2.2 TUNNELS

It is common practice in rock blasting to drill the holes parallel to a free face to which the explosives in the boreholes can break. In surface blasting work there are at least two free faces – one in front and one on the top surface.

In tunnelling, however, there is only one free face and that is the tunnel face itself. The advance is perpendicular to the tunnel face. The holes are drilled at right angles to the face so that the face itself cannot be used as a free face for breakage. A free face must be generated by a cut which will open a cavity into which the production holes in the round can break. This is achieved by a sequential blasting technique controlled by the delays of the detonators.

A satisfactory tunnel round requires a good cut along the entire length of the production holes.

The rock in a tunnel round is more effectively confined than in any other rock blast operation. The physical characteristics of the rock will consequently influence the result of the blast to a greater degree than in other work. The rock in small diameter tunnels is more firmly confined than in tunnels of larger diameter. The effect of this is that the specific consumption of explosive in tunnel work increases with decreasing cross section.

The most important factors which will influence the amount of drilling required are:
– Rock characteristics,
– Tunnel cross section,

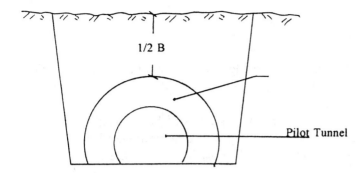

CROSS SECTION

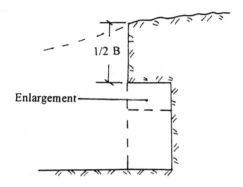

LONGITUDINAL SECTION

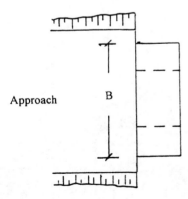

PLAN VIEW

Figure 2.1. The approach with pilot tunnel at the entrance.

- Borehole depth,
- Borehole diameter
- Type of cut used.

ANFO has to a great extent replaced cartridged explosives in tunnel work due to price, and quick and effective loading techniques with the loading equipment mounted on the drilling rig itself.

Under wet conditions cartridged explosives are still being used. The contour holes are normally loaded with column charges or detonating cord. Increased focusing on the environmental conditions will most likely increase the use of slurries in tunneling.

The use of electric detonators has for years been dominating in tunnel driving. For safety reasons non-electric detonating systems are rapidly gaining ground.

CHAPTER 3

Geology

3.1 GENERAL

The result of the blast is more dependent on the characteristics of the rock than on the explosives being used to break it.

The more important characteristics of the rock influencing the blasting result include:
– Tensile and compressive strength,
– Density
– Seismic velocity (acoustic velocity).

3.2 TENSILE AND COMPRESSIVE STRENGTH

Most types of rock have a compressive strength which is 8 to 10 times greater than the tensile strength.

These properties are important factors in rock blasting (Table 3.1).

3.3 DENSITY

Rock of high density is normally harder to blast than rock of lower density. One reason for this is that high density rock is heavier to move during detonation (Table 3.2).

3.4 SEISMIC VELOCITY

The seismic velocity (acoustic velocity) of the various types of rock varies from 1500-6000 m/s. Hard rock of high seismic velocity will shoot more easily, specially when explosives with high velocity of detonation are being used (Table 3.2).

Table 3.1.

Type of rock	Compressive strength (kg/cm^2)	Tensile strength (kg/cm^2)
Granites	2000-3600	100- 300
Diabase	2900-4000	190- 300
Marble	1500-1900	150- 250
Limestone	1300-2000	170- 300
Shale		300-1300
Sandstone, hard	3000	300

Table 3.2.

Type of rock	Density (kg/dm^3)	Seismic velocity (m/s)
Granite	2,7-2,8	4500-6000
Gneiss	2,5-2,6	4000-6000
Limestone	2,4-2,7	3000-4500
Dolomite	2,5-2,6	4500-5000
Sandstone	1,8-2,0	1500-2000
Clay mudstone	2,5-2,7	4000-5000
Marble	2,8-3,0	6000-7000
Diabase	2,8-3,1	4000-5000

3.5 ROCK STRUCTURE

The planning process should include a survey of the rock structure and other rock characteristics so that the drilling and loading pattern and direction of advance can be optimized as far as possible.

The rock structure included cracks and fissures and other zones of weakness.

Two expressions commonly used to describe the rock structure are 'strike' and 'dip'. Strike is the horizontal direction of the structure on the rock surface. Dip is the angle of the structure relative to the horizontal rock surface (Fig. 3.3).

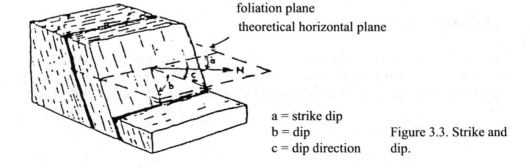

foliation plane
theoretical horizontal plane

a = strike dip
b = dip   Figure 3.3. Strike and
c = dip direction   dip.

When blasting in the dip direction the following can be expected:
Advantages: Good utilization of the energy of the explosives; good forward heave of the blast, easy to load; few problems with poor breakage at the toe.
Disadvantages: More back breakage and sliding; poor finished contours.

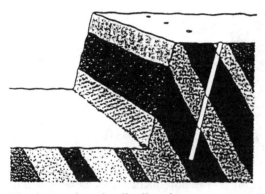

Blasting against the dip direction.
Advantages: Reduced back breakage.
Disadvantages: Less forward heave, more difficult to load and increased risk of poor breakage at the toe.

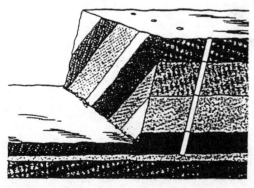

When blasting against the strike, poor results are achieved: Uneven floor; variable back breakage; uneven fragmentation.

The blast should be planned to achieve as favourable conditions as possible.

The foliation plane can be horizontal in limestone. The hole inclination can be varied to achieve better results. An advantage with this type of rock is that the amount of subdrilling can be reduced.

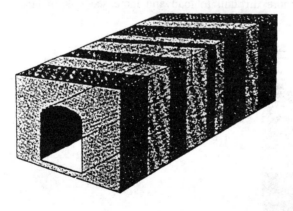

TUNNEL DRIVING
A strike at 90° to the direction of driving is normally favourable for good advance.

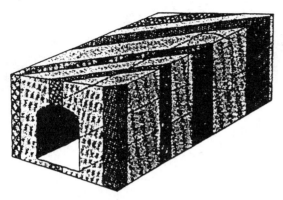

A strike at an angle to the direction of driving. The rock breaks more easily on the left side.

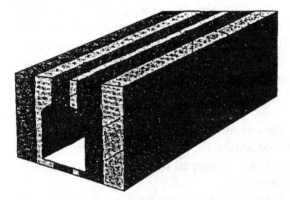

A strike parallel to the direction of driving. The result is poor advance and an uneven tunnel face.

CHAPTER 4

Drilling

4.1 GENERAL

In bench blasting the boreholes are drilled parallel to the free face against which the breakage can take place. In tunnelling, however, the only free face available is that of the tunnel face. Drilling parallel to a free face is consequently not possible. To achieve satisfactory blasting results an opening or void – the cut – must therefore first be created into which the surrounding rock can be blasted.

It is possible with carefully selected initiation systems and with delay detonators suitable for tunnel work to blast the cut and the production holes etc. in one operation.

Figure 4.1 shows expressions and nomenclature used in tunnel driving.

The creation of a proper cut is a precondition for a satisfactory tunnel blast. A number of different types of cuts have been developed over the years concurrent with the development of the drilling equipment.

The cuts can basically be divided into the following two groups:
1. V cuts: Wedge cuts and fan cuts,
2. Parallel cuts: Large diameter and burn cuts.
Parallel cut, large diameter, is the most commonly used cut. This type of cut is independent of the cross-sectional area of the tunnel and tunnel width. It is applicable over the entire cross section of the tunnel and for various borehole diameters.

A symmetrical wedge cut is also in use.

4.2 REQUIRED OFFSET

The offset is the deviation of the boreholes from a line parallel to the tunnel direction. The drilling equipment decides the necessary offset. Modern equipment requires an offset of 0.2-0.4 m. Lighter equipment requires less (Fig. 4.2).

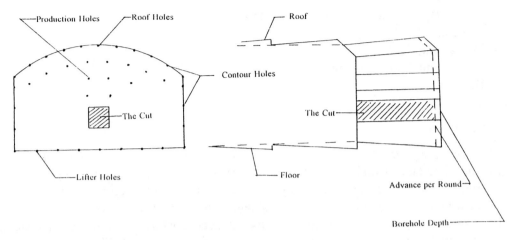

Figure 4.1. Nomenclature – tunnel driving.

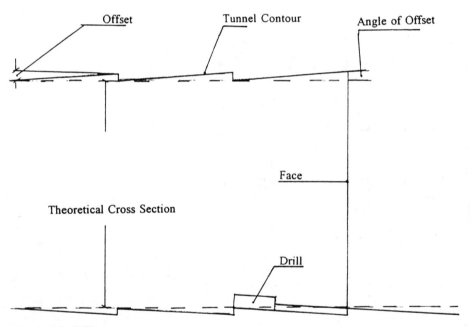

Figure 4.2. Offset.

4.3 TYPES OF CUT

4.3.1 *General*

The choice of cut is decided by the cross sectional area of the tunnel and its width, available drilling equipment, preferred advance and rock conditions.

The most commonly used types of cuts are:
– Fan cut,
– Wedge cut,
– Parallel cut, large diameter,
– Burn cut.

4.3.2 *Fan cut*

The principle of the fan cut is breakage/back ripping against the tunnel face, i.e. the only free face available. The fan cut is effective with respect to the total length of boreholes being drilled and quantity of explosives being used for each round. The cut gives a comparatively easy breakage, consuming less explosives. The fan cut requires a wide tunnel which limits the advance per blast. The non-symmetric drilling makes the cut less effective when applying modern drilling equipment. The fan cut is still being used in short tunnels and with plain drilling equipment (Fig. 4.3).

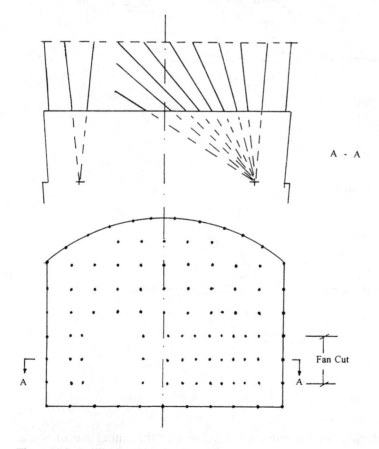

Figure 4.3. Drilling pattern for fan cut.

4.3.3 *Wedge cut*

In the wedge cut, the holes are drilled at an angle to the face in a symmetrical formation. The angle of the wedge must not be too narrow. The burden is increased with decreasing angle. The symmetric pattern distributes the drilling work evenly between the drills. The increased burden requires more explosives and more drilling work as compared with the fan cut. The advance depends upon the tunnel width and the wedge cut is consequently most suited in wider tunnels.

A wedge cut blast throws the broken rock further out of the tunnel as compared with other types of cuts. This makes it more cumbersome to scale the roof and the face from a position on top of the muckpile and it increases mucking time. The greater throw of fired rock is also more likely to cause damage to ventilation ducting close to the face (Fig. 4.4).

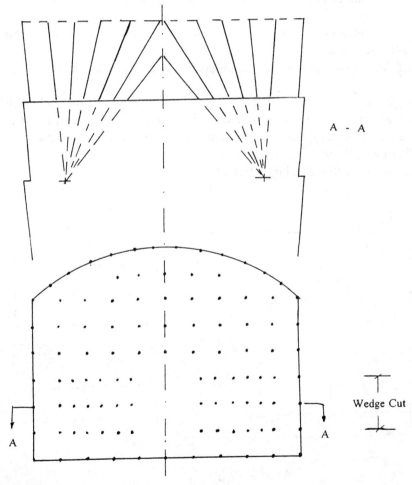

A - A

Wedge Cut

Figure 4.4. Drilling pattern for wedge cut.

4.3.4 *Parallel cut*

The parallel cut consists of one or more larger diameter unloaded boreholes. All holes are drilled at a right angle to the face and parallel to the tunnel direction. The breakage is against the opening or void formed by these unloaded holes of diameter 76-150 mm.

This opening in gradually opened up by successive detonation of the adjacent loaded holes and the pulverized rock is blown out of the cut. The parallel cut requires a weak load along its entire length. Packaged explosives of a composition giving gas energy with low gas temperatures are used.

ANFO is also, to a large extent being used in the parallel cut procedure. The use of explosives with high speed of detonation may easily cause sintering, reduced blowout and an unsuccessful blast.

Modern electric/hydraulic drilling equipment will drill up to 152 mm in diameter. The large diameter cut holes are initially drilled as pilot holes, and of the same diameter as the production holes.

Several large diameter cut holes will help to secure good breakage along the entire length of the advance.

The length of the advance is in principle independent of the cross section of the tunnel.

Parallel cuts give in general better breakage and fragmentation of the rock with less throw-out and spreading as compared with the V cuts. This in turn reduces loading time and also makes it more convenient to scale the roof, walls and face of loose rock from a position on top of the pile.

Figure 4.5 shows a parallel cut, large diameter.

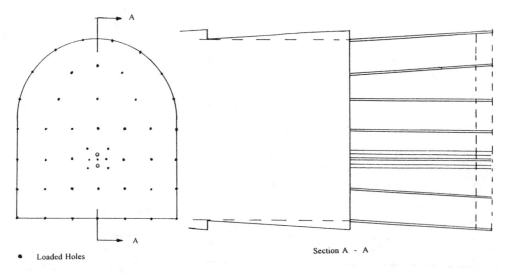

● Loaded Holes

○ Empty Large Diameter Holes

Section A - A

Figure 4.5. Drilling pattern – parallel cut, large diameter.

Bedding and blastability of the rock together with borehole diameter and length of advance are important factors in the planning process.

The advantages of the parallel cut, large diameter, are:
– Well suited for drilling with modern drilling equipment,
– Well suited for rounds of long length,
– The length of the advance is in principle independent of the cross section or width of the tunnel,
– Good breakage,
– Moderate throw-out and spreading of the pile. Reduced loading time and scaling of roof, walls and face from a position on top of the pile,
– Good fragmentation.

It is a precondition of the parallel cut, large diameter, that the rock blasted into the empty boreholes has sufficient room for complete blowout. Based on experience, this requires very accurate drilling and carefully selected delay times and sequence of detonation for the loaded holes adjacent to the cut (Figs 4.6, 4.7, 4.8, 4.9 and 4.10).

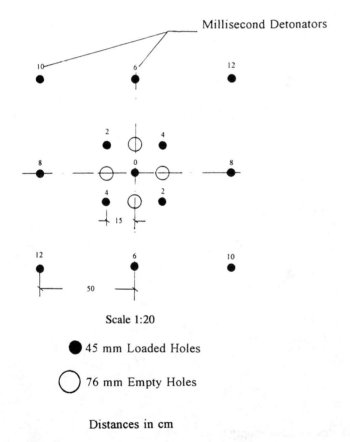

Figure 4.6. Parallel cut – large diameter.

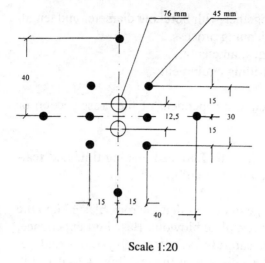

Scale 1:20

Distances in cm

Figure 4.7. Parallel cut – large diameter with two empty holes (76 mm).

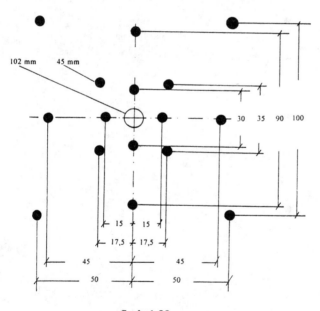

Scale 1:20

Distances in cm

Figure 4.8. Parallel cut – large holes with one empty hole.

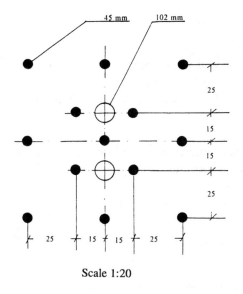

Scale 1:20

Distances in cm

Figure 4.9. Parallel cut – large diameter with two empty holes.

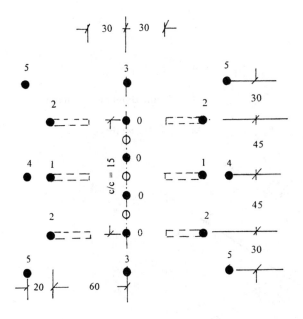

○ Empty Holes, 34 - 37 mm diameter   ● Loaded Holes, 34 - 37

Distances in cm

Figure 4.10. Burn cut/wedge cut.

4.3.5 *Blast design for parallel cut, large diameter*

The following conditions must be satisfied:
− Sufficient opening for the rock to be blasted and blown out,
− The burden must be in the right proportion to the opening.
When designing the cut, use curve Figure 4.11 as a starting point to find the required large diameter hole area for the cut.
    The curve is based on experience.
    The distance (burden V) between the large diameter unloaded cut holes and the adjacent loaded holes is given by the following formula:

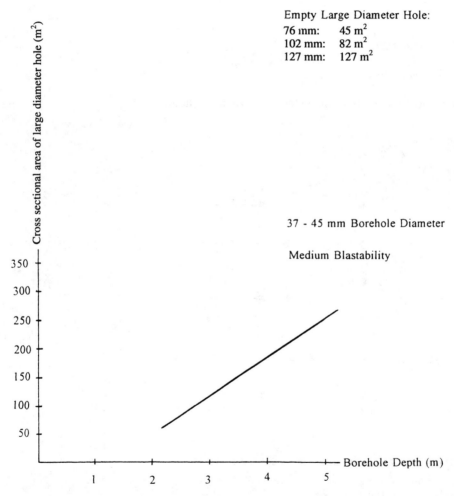

Figure 4.11. Required area of empty large diameter holes for parallel cut as a function of borehole depth, blastability and borehole diameter.

Burden $V = 1.5\text{-}2.0 \times d$

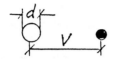

● Loaded hole: Diameter 37-45 mm
○ Large diameter hole: Diameter 76-102-127 mm
Other dimensions in meters.

The burden V for the other holes of the cut is governed by the width B of the established opening. Based on experience with rock of medium blastability:

Burden $V = 0.7 \times B$

Figure 4.12 gives suggested burden as a function of the established opening.

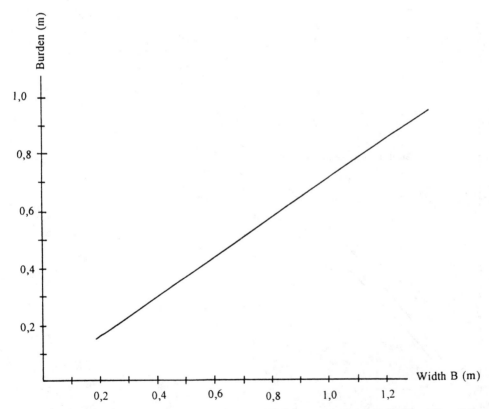

Figure 4.12. Suggested burden as a function of blasted opening. Rock of medium blastability.

4.4 REQUIRED DRILLING

The required amount of drilling is specified as the number of boreholes per round for the given cross-section of the tunnel.

The more important factors governing the extent of required drilling include:
– Blastability of the rock,
– Cross section of the tunnel,
– Borehole depth,
– Borehole diameter,
– Type of cut to be used.

For rail and road tunnels and rock caverns specifications are often given for the contour blasting in terms of boreholes distances, type and quantity of explosives being used.

The required number of boreholes, exclusive of the large diameter unloaded holes of the parallel cut, is given in Figure 4.13 for 34 and 45 mm diameter holes.

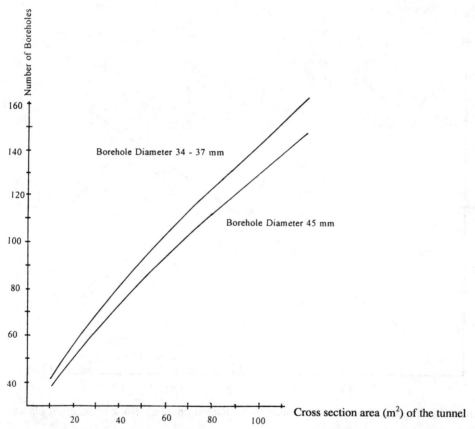

Figure 4.13. Required number of boreholes exclusive of the large diameter holes in the parallel cut – rock of medium blastability – borehole depth 4.5-5.0 m.

4.5 DRILLING PATTERN

Start the drilling pattern design by positioning the cut and the contour holes first. Secondly, get the production holes adjacent to the contour in place and fix the position of the lifter holes. Finally, position the production holes. When fixing the production holes adjacent to the cut make sure to allow sufficient room for the cut breakage.

The offset is between 0.2-0.4 m. The burden in the transition zone between the contour holes and the production holes must be reduced by an amount equal to the offset.

For practical reasons, to ease the drilling and loading operations, it is normal to position the cut 1.0-1.5 m up from the floor and alternately on either side of the centerline to avoid collating a hole into a socket.

It is important to facilitate effective drilling when positioning the cut and designing the drilling pattern. An even distribution of the drilling work between the drills is necessary for making the drilling time as short as possible. The loading of the cut is most effectively done from the floor.

Blasting as close as possible to the pre-fixed profile is an ever increasing requirement – likewise with respect to the quality of the remaining rock. Accurate drilling of the contour holes is important in this respect. Too much offset of the contour holes will give a pronounced 'sawtooth' shaped profile and extensive damage to the remaining rock in the transition zone between the blasts.

The holes adjacent to the contour holes shall be given an offset angle equal to that of the contour holes. Poorly performed contour blasting will give an unnecessary amount of back breakage, increased clean up and stabilisation work, and poor quality of the final product. The result will be loss of pressure in water tunnels and increased maintenance for rail and road tunnels. The risk of falling rock is dangerous and also a traffic hazard.

The lifter holes should be given the same offset angle as that of the contour holes.

$$\text{Number of lifter holes} = \frac{\text{Tunnel width} + 2 \times \text{offset of contour holes} + 1}{\text{Distance between holes}}$$

Suggested values for burden and distances between holes based on practical experience and rock of medium blastability:

Borehole diameter: 34-38 mm:

Contour holes:	Burden 0.6-0.8 mm	Distance between holes 0.5-0.8 m
Next row:	Burden 0.9	Distance between holes 0.8 m
Lifter holes:	Burden 0.8 m	Distance between holes 0.7 m

Borehole diameter: 45 mm:

Contour holes:	Burden 0.8-1.0 m	Distance between holes 0.7-1.0 m
Next row:	Burden 1.0 m	Distance between holes 1.1 m
Lifter holes:	Burden 1.0 m	Distance between holes 1.0 m

Figures 4.14 and 4.15 give examples of drilling patterns for 37 and 45 mm borehole diameters, respectively.

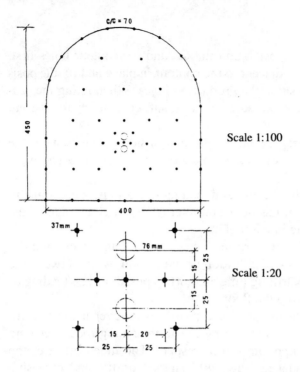

Scale 1:100

Scale 1:20

Tunnel Cross Section Area	16 m²
Borehole Diameter: Loaded Holes	37 mm
Empty Holes	76 mm
Number of Holes:	47 + 2 (Empty)
Borehole Depth	3,5 m

Figure 4.14. Example of drilling pattern for 37 mm borehole diameter.

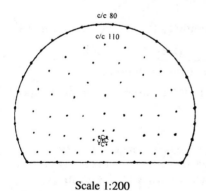

Scale 1:200

Tunnel Cross Section Area	60 m²
Number of Holes with Diameter 45 mm	93
Number of Holes with Diameter 102 mm	4
Borehole Depth	5 m
Medium Blastability	

Figure 4.15. Example of drilling pattern for 45 mm borehole diameter.

CHAPTER 5

Loading

5.1 GENERAL

Originally, tunnels and rock caverns were almost exclusively blasted with explosives in packaged form. Lately, the packaged explosives have to a large extent been substituted by bulk ANFO in tunnels using borehole diameters of 45 mm and larger. Modern loading equipment and better quality ANFO have made bulk ANFO competitive both in terms of blasting efficiency and price.

5.2 LOADING WITH PACKAGED EXPLOSIVES

The fan cut and the wedge cut are almost universally loaded with dynamite and well tamped to achieve a high loading density.

The holes of the parallel cut adjacent to the unloaded center holes are loaded with a low energy explosive to prevent sintering and reduced blow-out. Use an explosive with relatively low gas temperature.

The production holes are loaded with a high energy explosive for the bottom charge and with a weaker explosive in the rest of the hole.

The lifter holes are loaded with dynamite partly because of the generally wet conditions at floor level and partly to produce a muckpile which is easy to handle.

The contour holes are loaded with rigid column charges (plastic tubes) of 17 mm and 22 mm in diameter and with a dynamite cartridge at the bottom of the holes as an initiator. The rigid column charges are fitted with a specially designed collar to center the charge in the hole and leave a buffer of air around it. This will reduce the fracturing of the rock in the contour zone.

5.3 LOADING WITH ANFO

ANFO gives a higher loading density than cartridged explosives, primarily because the ANFO fills the bore hole more completely. This in turn leads to a higher specific consumption of explosives with bulk ANFO as compared with the use of cartridge explosives.

A half or full length of dynamite cartridge is used as a bottom charge and primer in boreholes loaded with ANFO.

The pneumatic ANFO-loader is either mounted on the drilling rig itself or on a separate unit. A pressure loading system is mostly used and with two or three outlets for the loading hoses. The pressure chamber contains from 150 to 500 kg explosives according to requirement.

Blowing, de-watering and loading is done with the same hose.

The loading process is remotely controlled. In addition to the start/stop function, the blowing and the de-watering can also be remotely operated.

It has gradually become quite common under dry conditions to load both the cut, the production holes and partly the contour holes with ANFO and a primer.

5.4 LOADING OF THE CONTOUR HOLES

Detonating cord of 80 and 40 gr m strength have shown to be a good explosive for the contour. The result will vary with the condition of the rock, distance between the holes and the burden at the contour. The optimal combination must be derived at by testing in each separate case.

ANFO can be used in the contour holes after fitting a nozzle at the end of the hose.

The loading density of the contour holes is normally approx. 25% of the normal density. The amount is controlled by the loading pressure, rate of retracting the hose, the width of the split and the form of the nozzle at the end of the hose (shaped like a whistle).

5.5 SUGGESTED LOADING FIGURES

Required consumption of explosives is given in kilogramme per cubic meter of rock (kg/m^3) for the cross section of the tunnel. The expression 'specific explosive consumption' will also be used.

Suggested loading is given by Figures 5.1-5.3 as functions of the blastability of the rock and cross section of the tunnel for borehole diameters 34-37 mm and 45 mm and parallel cut, large diameter.

Figure 5.1 is applicable for 34-37 mm borehole diameter, parallel cut, large diameter and cartridged explosives. Rock of medium blastability.

Figure 5.2 is applicable for 45 mm borehole diameter, parallel cut, large diameter and cartridged explosives. Rock of medium blastability.

Figure 5.3 is applicable for 45 mm borehole diameter, parallel cut, large diameter and ANFO. Rock of medium blastability.

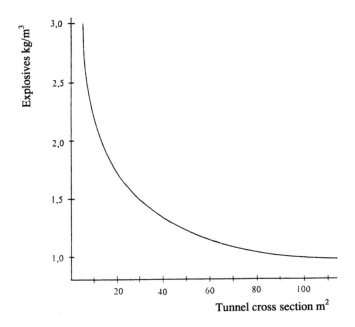

Figure 5.1. Suggested loading for 34-37 mm borehole diameter; parallel cut, large diameter and cartridge explosives – medium blastable rock.

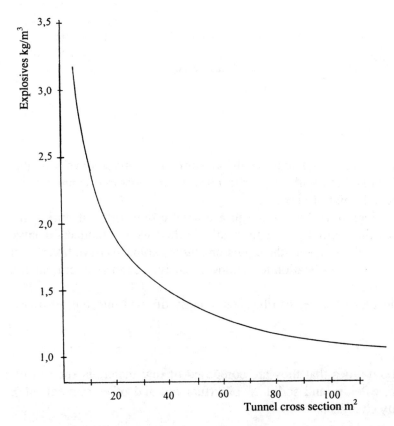

Figure 5.2. Suggested loading for 45 mm borehole diameter, parallel cut, large diameter and cartridged explosives – medium blastable rock.

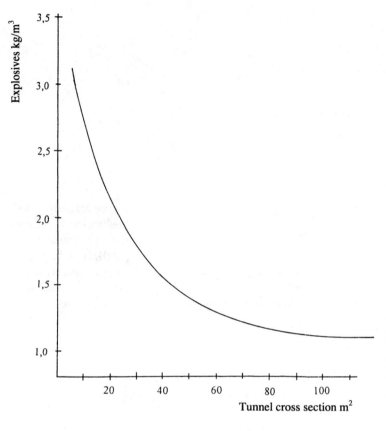

Figure 5.3. Suggested loading with ANFO and 45 mm borehole diameter, parallel cut, large diameter and ANFO medium blastable rock.

5.6 EMULSION FOR TUNNELS

It has lately been demonstrated that emulsion slurry explosives have valuable qualities in tunnel work both with respect to blasting and loading technique and from an environmental point of view.

The slurry explosives have been through a dramatic development during the last 30 years. When they initially became available they were intended for large boreholes in open pit mines. Today the emulsion slurry explosves have taken over and enlarged their field of application to include extensive use in smaller quarries, road cuttings and tunnels.

The slurry explosives of today can be most readily divided into the following two types:

1. Watergels,
2. Emulsions.

It is common to both types that they are composed of raw materials such as nitrates (ammonium, calcium and sodium), oils (fuel oil, and other mineral oils), water and speciality chemicals.

In simple terms the difference between the two types is that watergel has very small droplets of fuel oil suspended in a nitrate solution whilst emulsion has microscopically fine droplets of nitrate solution suspended and surrounded by a continuous oil phase.

The contact between the fuel oil and the nitrate is much better in an emulsion which gives a more complete chemical reaction with more energy liberated, higher speed of detonation and less emission of noxious gases which is an important factor when working under ground.

Both watergels and emulsions have a high degree of water resistance.

Another important factor is the higher level of safety in us, transport and during production compared with more conventional explosives.

Optimal safety is achieved when using the site mixed or the site sensitized systems. The emulsion is a mixing together of components which are not by themselves defined as explosives. The mixture becomes an explosive during the loading process when the density is reduced by adding a gasing agent.

Pumpable emulsion explosives have been tried in tunnels with good results in the last few years.

On the positive side:
– Reduced emission of noxious gases,
– A higher level of safety,
– Increased water resistance,
– The same type of explosives can be used in all holes.

On the negative side:
– The equipment is more complicated and requires trained operators.

Emulsions are water resistant and pumpable explosives. This makes it possible to vary the quantity to be loaded from one hole to the next. The lifters and the production holes need a maximum of explosive energy whilst the contour holes require considerably reduced loading. This is achieved by mechanically operated withdrawal of the loading hose. The hose is pulled back at a certain rate while the pump is delivering an even amount of emulsion per unit of time. The parameters are adjusted to give the required amount of emulsion per meter of borehole.

Safety is optimized by gasing the emulsion during loading.

Storage and transport hazards of the explosive are eliminated (see Appendix 4).

CHAPTER 6

Detonators – firing pattern

6.1 GENERAL

95% of the detonators used underground in Scandinavia are of the non-electric type, only 5% are electric.

The non-electric detonators are primarily the Nonel and similar products. They represent a flexible system and can be used in almost any type of blasting work.

Electric detonators are available in 3 different groups classified according to sensitivity. Detonators in Group 3 have the highest protection against accidental firing by extraneous electricity. They are mostly used in tunnels when loading from electric/hydraulic drilling rigs.

Electric detonators, subdivided in 3 groups, have the following electrical characteristics:

Characteristics	Group 1	Group 2	Group 3
Initiation impuls mJ/ohm	2.5-5.5	80-140	110-2500
Minimum current (amp)	0.28	1.20	4.0
Critical current for circuit (amp)	1.1	3.5	25.0
Water resistance	24 hours under pressure corresponding to a water head of 30 m		

The firing pattern must be designed so that each borehole or group of boreholes are given as good break-out conditions as possible and a minimum of burden.

It is also important that the rock to be blasted at each delay interval has sufficient room for break-out.

6.2 THE CUT

The delay intervals in the cut must be sufficient for the rock to be blown out of the hole before the next hole is fired. The delay time should be between 50 and 100 milliseconds. Soft rocks require longer delays than hard rock. The millisecond detonators have a delay time of 25-30 milliseconds. This means that only every second or third delay number shall be used. See Figures 6.1 and 6.2.

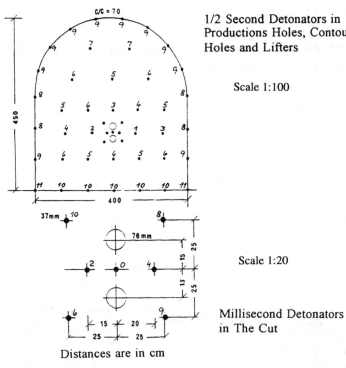

1/2 Second Detonators in Productions Holes, Contour Holes and Lifters

Scale 1:100

Scale 1:20

Millisecond Detonators in The Cut

Distances are in cm

Figure 6.1. Example of firing pattern with electric detonators, millisecond and ½ second detonators – tunnel cross section 16 m².

ELECTRIC DETONATORS

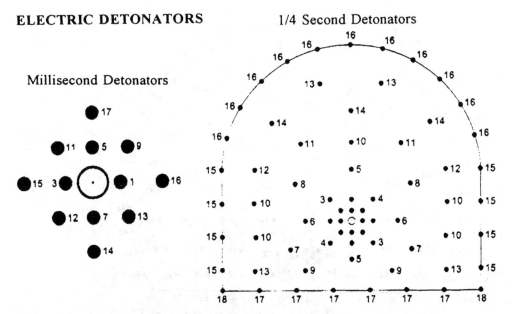

Figure 6.2. Example of firing pattern with millisecond and ¼ second detonators – tunnel cross section 30 m² borehole diameter 45 mm.

6.3 THE PRODUCTION HOLES

The firing pattern for the production holes shall be arranged to give the best possible break for each separate hole. See Figures 6.1-6.4.

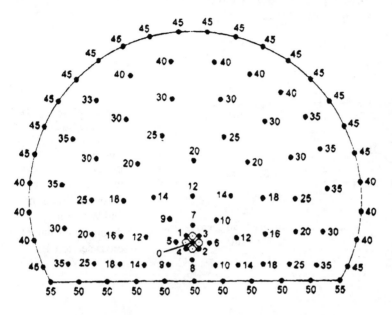

Figure 6.3. Example of firing pattern with nonel LP. Tunnel cross section 60 m² borehole diameter 45 mm.

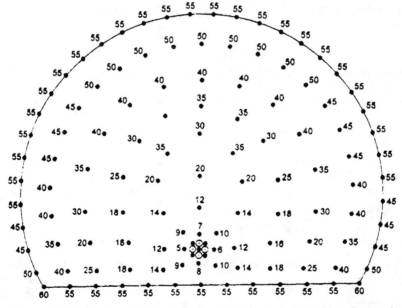

Figure 6.4. Example of firing pattern with Nonel LP tunnel cross section 90 m² borehole diameter 45 mm.

6.4 THE CONTOUR HOLES

The contour holes along the wall are fired with the same delay number to achieve as much interaction as possible between charges. This is referred to as smooth blasting of the contour or trim blasting. The same applies to the contour holes in the roof which are fires with a higher delay number than the contour holes along the wall. See Figures 6.1-6.4.

6.5 THE LIFTERS

The lifter holes are fired with later delay numbers than the holes in the roof. See Figures 6.1-6.4.

6.6 NONEL LP IN TUNNEL WORK

NONEL LP is a detonator specially designed for blasting and other underground work. It is important in tunnel blasting that the intervals used are sufficiently long for the broken rock to be heaved away from the face before the next holes are detonated.

Soft rock and rock hard to blast normally require longer delay time between intervals than harder rock and rock which is more easily blasted.

Interval times for NONEL LP.

Interval number	Delay time milliseconds	Interval time milliseconds
0	25	
1	100	75
2	200	100
3	300	100
4	400	100
5	500	100
6	600	100
7	700	100
8	800	100
9	900	100
10	1000	100
11	1110	110
12	1235	125
14	1400	165
16	1600	200
18	1800	200
20	2075	275
25	2500	425
30	3000	500
35	3500	500
40	4000	500
45	4500	500
50	5000	500
55	5500	500
60	6000	500

6.6.1 *Alternative hook-ups*

Nonel Bunch Connectors – Figure 6.5
The hook-up of NONEL LP in tunnelling is quickest and most easily done with what is known as bunch connectors. For this method special hook-up units, Nonel Bunch Connectors are available.

Nonel tubings from the loaded boreholes are collected in bunches with max. 20 tubings per bunch. A bunch connector is fastened in each bunch by the affixed loop of detonating cord. Then the tubings of the bunch connectors are coupled together in a snapline 0 and the blast is ready for firing.

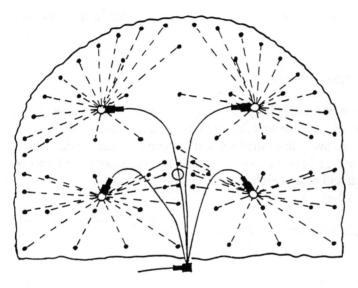

Figure 6.5. Connection of tunnel round with Nonel Bunch Connectors.

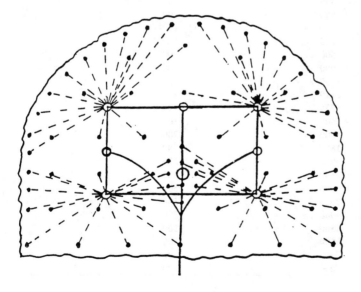

Figure 6.6. The bunches tied up and connected together with detonating cord – 5 gr/m.

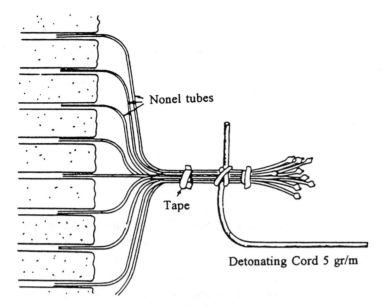

Nonel tubes

Tape

Detonating Cord 5 gr/m

Figure 6.7. Bunch ig-
nition.

Hook-up with detonating cord – Figure 6.6
Detonating cord can be used for bunch ignition as an alternative to the use of
Nonel Bunch Connectors. The bunches are hooked together with detonating cord.
Each bunch is knotted together with a half hitch.

This method requires precision to prevent the shock pressure from the detonating cord to tear loose the nonel tubing without ignition. For this method, detonating cord of strength 3.6, alternatively 5.0 or 7.0 g/meter is used.
– With bunch ignition, the nonel tubing should be approx. 2 meter longer than
   the borehole dept (Fig. 6.7),
– The distance between a detonating cord and a parallel nonel tubing should be at
   least 0.1 meter (Fig. 6.7).

CHAPTER 7

Safety procedures

7.1 GENERAL

In shaft and tunnel work a number of boreholes are drilled in highly constricted areas. Under such circumstances the risk of drilling into boreholes with explosives is higher than in other type of construction work. It is important to inspect the tunnel face carefully and look for undetonated explosives from the last round and make sure that the drill is not placed in a borehole of the previous blast. This should be routine work performed according to a fixed procedure.

The possibility of encountering inflammable gases in the rock structure under ground must be considered when specifying the explosives, detonators and other blasting accessories to be used (Fig. 7.1).

7.2 PROCEDURE FOR DRILLING, LOADING AND BLASTING IN TUNNELS

7.2.1 *Drilling*

The face shall be carefully cleared of loose rock, hosed down with water, secured and cleaned before the start of drilling.

Never drill into a borehole or socket which contains, have contained or may have contained explosives.

New boreholes shall be positioned so that previous boreholes can not be intercepted.

Simultaneous drilling and loading should be avoided (Figs 7.2 and 7.3).

The distance between a hole being drilled and a hole partly or fully loaded shall under no circumstances be less than 2 meters.

Nobody shall be in front of the drilling rig during operation.

The face shall be carefully inspected for left-overs of explosives before new drilling starts. Such inspections shall be done by washing down with water.

Boreholes with undetonated or risk of undetonated explosives shall be clearly marked.

Figure 7.1. Safety procedures. No! No! No! No! No! and absolutely No!!!

Figure 7.2. IT is prohibited to load during the drilling operation.

Drilling must not take place above holes being loaded. Loose rock may fall down on explosives and detonators.

Drilling rods which are clogged shall not be used. They must be carefully marked and not kept together with the rest of the rods. De-clogging of the drilling rods may be tried with the use of water under pressure.

Figure 7.3. This is prohibited!

7.2.2 *Loading*

The initiating cartridge or primer shall be inserted first and the detonator shall point in the direction of the charge.

7.2.3 *Blasting*

Immediately after the blast, check that the round fired normally before proceeding further.

Undetonated explosives shall be taken care of immediately by placing and detonating a new charge in close contact with the undetonated one.

If the shooting of undetonated explosives is unsuccessful, try to remove it with water under pressure.

After undetonated explosives have been taken care of, the area shall be carefully inspected again.

7.3 ELECTRIC CURRENT LEAKAGE

7.3.1   *Current leakage from electric installations*

Electric installations and electric equipment can have faulty insulation resulting in stray currents and voltage potentials at different points in the rock or between the ground and metallic equipment (rails, drilling rig, tubing etc.). If uninsulated parts of the leading wires of an electric detonator come in contact with points of different potentials, a current can be set up which may initiate the detonator inadvertently (Fig. 7.4).

7.3.2   *Safety rules*

During loading with electric detonators the following points must be observed:

Do not let the leads to the detonators, at any time during loading, come in contact with foreign metallic objects (Fig. 7.4).

Do not place electric lamps and electric pumps too close to the work place during loading (Fig. 7.4).

Figure 7.4. Electric equipment must be in good working order... and without electric current leakages.

7.3.3   *Insulation of bared ends of the leads*

Increased safety during loading is achieved by the use of insulator sleeves at the free end of the leads.

7.4 HIGH VOLTAGE OVERHEAD WIRES AND CABLES

When blasting close to high voltage overhead lines, consideration must be given both to the electric-magnetic field (induction) and possible stray currents. See Figure 7.5.

7.4.1   *Closed circuits – conductive contact*

Precautions must be taken so that the shot firing circuits do not make conductive contact with the ground thus forming a closed circuit capable of creating a current generated by induction. See Figure 7.5.

7.4.2   *Insulation – blasting mats*

In practice, use insulating sleeves or insulating tape to prevent the joints from making conductive contact with the ground or metallic objects. Do not short circuit the leading wires and keep the free ends separately insulated. The blasting-mats and leading wires should be secured to the ground to prevent them from being thrown up into overhead wires and exposing the shot firing operator to the hazard of dangerous electric shocks. See Figure 7.5.

7.4.3   *Detonators with low sensitivity or non-electric detonators*

Increased safety is achieved by using detonators with low sensitivity. Do not

Figure 7.5. Be on the alert! Uninsulated parts of the electric leads must not come in contact with the ground or conductive objects.

shorten the leading wires during hook-up. A shortening of the leading wires reduces the electric resistance of the detonator and increases its sensitivity.

Under unfavourable conditions it should be considered using non-electric initiation systems. See Table 7.1 for safety distances.

7.5 RADIO ENERGY

7.5.1 *The radio aerial effect*

It is a small risk of electric detonators being inadvertently fired by radio waves. Nevertheless, it is possible under special conditions. The detonator circuit can act as an aerial and pick up radio waves from the sender. The induced current will increase with decreasing distance between the radio transmitter and the circuit.

7.5.2 *Induction*

The current generated by induction can vary in strength along the circuit. Detonators positioned in the circuit where the induced current reaches a maximum can inadvertently fire. Induced current will not be generated if the detonators are closely packed together as delivered from the supplier.

Table 7.1. Safety distances. When working close to high voltage power lines, the following safety distances must be observed to prevent inadvertent firing of detonators.

Group 1 detonators Grid Voltage (kV)	Overhead lines, distance (m)	Cables in the ground, distance (m)
0.4-6	20	2
7-12	50	3
13-24	70	6
25-52	100	10
> 52	200	16
Group 2 and Group 3 detonators Grid Voltage (kV)	Overhead lines, distance (m)	Cables in the ground, distance (m)
> 24	5	2
25-72.5	6	3
72.6-123	10	10
124-245	12	10
> 245	16	16

Note: The distances are plan distances except for loading under ground when the distances are measured on the inclination; use non-electric initiation systems when the distances to the overhead lines are less than that given in the table; no restrictions apply when the grid voltage is less than 0.4 kV.

Table 7.2. Blasting close to radio transmitters. The following distances must be respected when shooting with electric detonators of Group 1 (regular electric detonators) close to radio transmitters.

Transmitter power (watts)	Distance (m)	Transmitted power (Kw)	Distance (m)
5	4	1	40
10	10	4	75
50	15	10	95
100	20	50	150
200	25	100	200
300	30	200	250
500	35	300	300
		500	350
		750	400
		2000	650

If the blasting work takes place adjacent to a radio transmitter, the safety distances applicable for detonators of Group 1, as shown in Table 7.2, shall be applied.

7.5.3 *Safety rules*

The electronic detonator circuit must not come in conductive contact with the ground. Use insulation sleeves and insulation tape. The circuit shall rest on the ground or as close to the ground as possible. It is important that no part of the detonator circuit sticks up above the rest of the circuit.

The electric leads must not be shortened when connecting. Leads which are too long, shall be rolled up and stuck into the borehole. It is recommended to use low sensitivity detonators or non-electric detonators when shooting close to a radio transmitter station.

7.6 THUNDERSTORMS

Under thundery weather and risk of lightning, the procedure is to discontinue all loading with electric detonators (Fig. 7.6).

7.6.1 *Blasting under ground*

The same procedure as above apply when blasting work is in progress deep under ground. A strike of lightning can penetrate deep down into the ground or follow rails and piping with negligible drop in voltage potential.

Figure 7.6. Under thundery weather and risk of lightening, the procedure is to discontinue loading with electric detonators.

7.6.2 *Lightning warning devices*

Effective lightning warning devices are available. They register flashes of lightning and electricity in the atmosphere.

Non-electric detonators should be used on a permanent basis in exposed areas.

7.7 STATIC ELECTRICITY

Static electricity can under special conditions be built up in air of low relative humidity. The air jets from air operated equipment can whirl up and send dry rock dust against metallic and plastic objects which are not properly earthed and build up static electricity. If a detonator makes contact with these objects, it may fire. Static electricity can also be built up by blowing material in powder form, like ammonium nitrate, through hoses which have not been treated to prevent build-up of static electricity.

Exhaust emission from vehicles can also cause static electricity build-up and should not be allowed too close to places where detonators are being hooked up (Fig. 7.3).

CHAPTER 8

Tunnel blast design illustrations

8.1 A 16 M² TUNNEL CROSS SECTION – CARTRIDGE EXPLOSIVES

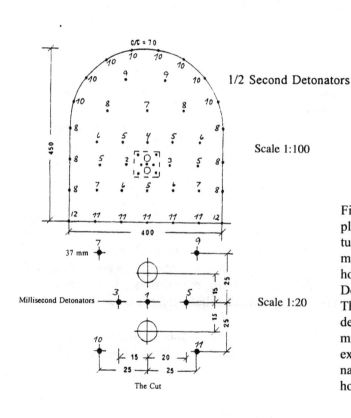

1/2 Second Detonators

Scale 1:100

Millisecond Detonators

Scale 1:20

The Cut

Figure 8.1. Drilling and firing plan for 16 m² cross section tunnel. Number of holes of 37 mm diameter: 47. Number of holes of 102 mm diameter: 2. Detonators: Millisecond in The Cut, otherwise ½ second detonators. Explosives: 32 mm and 25 mm cartridged explosives and 80 gr/m detonating cord in the contour holes.

The following criteria govern the design of the drilling, loading and firing plan:
– Cross section of tunnel   16 m²
– Blastability   Medium
– Borehole diameter   37 mm
– Borehole depth   3,5 m
– Contour   Smooth blasting
Parallel cut, large diameter, area   150 cm² (curve Fig. 4.11)

42

Number of large diameter holes   2 @ 102 mm – 164 cm^2
Distance between large diameter hole and first loaded hole: $a = 1.5 \times 10.2$ mm = 15 cm
Number of holes of 37 mm diameter   47 (curve Fig. 4.13)
Explosives per m^3 of rock (specific loading)   1,85 kg/m^3 (curve Fig. 5.1)
Rock per round: 16 m^2 × 3.5 m =   56 m^3
Drilled meters per m^3: $(47 + 2) \times 3.5/56$ m^3) =   3,06 m/m^3
Explosives per round total: 56 m^3 × 1.85 kg/m^3 =   103,6 kg
Cut/production/lifter/wall: Cartridged explosives of diameter 32 and 25 mm
Contour (roof): One cartridge of 25 mm and 80 gr/m detonating cord
Detonators: Millisecond (Group 3) in the cut, otherwise 1/2 – second (Group 3)

8.2 A 30 M^2 TUNNEL CROSS SECTION – CARTRIDGED EXPLOSIVES

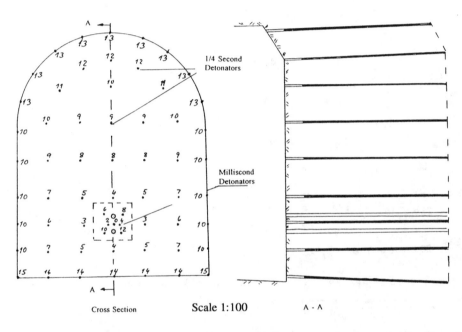

Figure 8.2. Drilling and firing plan for a 30 m^2 cross section tunnel. Number of 45 mm holes: 63. Number of 102 mm holes: 2.

The following assumptions have been made when designing the drilling-, loading- and firing plan:
– Cross section of tunnel   30 m^2
– Blastability   Medium
– Borehole diameter   45 mm
– Borehole depth   4.3 M
PARALLEL CUT, LARGE DIAMETER, AREA   200 cm^2 (curve Fig. 4.11)
Number of large diameter holes of 127 mm diameter   2 (250 cm^2) (Fig. 4.11)

Distance between large diameter hole and first loaded hole: $a = 1.5 \times 12.7 = 20$ cm

Number of holels of 45 mm diameter	63 (curve Fig. 4.13)
Explosives per m³ rock:	1.65 kg/m³ (curve Fig. 5.2)
Rock per round 30 m² × 4.3 m =	129 m²
Drilled meters per m²: (63 + 2) × 4.3 m/129 m³ =	2.17 m/m³
Explosives per round total: 1.65 kg/m³ × 129 m³ =	213 kg

Cut/production/lifter/wall: Cartridged explosives of diameter 32-35 mm
Contour (roof): One cartridge of 32 mm diameter and detonating cord, 80 gr/m
Detonators: Millisecond in the cut, otherwise ¼ – second (Group 3)

8.3 A 60 M² TUNNEL CROSS SECTION – BLASTING WITH ANFO

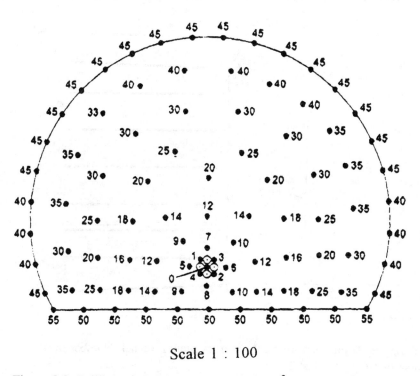

Scale 1 : 100

Figure 8.3. Drilling- and firing plant for a 60 m² tunnel. Number of 45 mm holes: 93. Number of 102 mm holes: 4. Borehole depth: 4.3 m. Detonators: NONEL – LP, No. 0-55.

The following assumptions have been made when designing the drilling, loading and firing plant:

– Cross section of tunnel	60 m²
– Blastability	Medium
– Borehole diameter	45 mm
– Borehole depth	4.3 m

No problems with water at the tunnel face. ANFO to be used as the main explosive and with non-electric detonators

Parallel cut, large diameter with 4 large diameter holes of 102 mm diameter to be used. Total parallel cut area = 328 cm^2

Distance between large diameter hole and first loaded hole is $a = 1.75 \times 10.2 = 17.9$ cm. Adopt $a = 20$ cm

Required number of boreholes of 45 mm diameter: 93 (curve Fig. 4.13)

Loading plan:

The cut	ANFO
Lifter holes:	ANFO alternatively 32 mm cartridges
Production holes:	ANFO
Contour:	Detonating cord, 80 gr/m
	Use one cartridge of 32 mm as a primer or starter in each hole
Total explosives per m^3:	1.4 kg/m^3 (curve Fig. 5.3). Adopt 1.45 kg/m^3 to allow for waist during loading
Explosives, total:	60 m$^2 \times 4.3$ m $\times 1.45$ kg/m^3 = 374.1 kg
Detonators:	NONEL – LP

CHAPTER 9

Tunnels and tunnel break-through under water

9.1 GENERAL

It is quite common in a number of countries to have deep inland lakes located high up in the mountains. These lakes are widely used as reservoirs for the production of electric power and tapped below their normal levels. The tapping is done through a tunnel system driven into the lake at a predetermined depth. The tapping is controlled by gates and hatches (Fig. 9.1).

Most tunnels and tunnel break-through into inland lakes have been made in order to use the lake as a water reservoir, primarily for electric power generation and drinking water supply.

Sub-sea tunnels are also used to bring oil and gas ashore from offshore fields.

The general criteria concerning the location of the break-through will be the same regardless of purpose and depth.

The deepest break-through registered for power generation is at a depth of 120 m (1987 in Norway).

Three break-through at the end of sub-sea tunnels several km long for oil and gas have been made at 180, 80 and 60 m depth, respectively. They have been made in Norway in connection with oil and gas activities in the North Sea.

9.2 PRELIMINARY SURVEYS

Sediments are common in lakes and at the sea bed, but do not necessarily create a problem.

Small amounts of loose sediments which are stable and not apt to slide, can be an advantage above the break-through. These sediments will fill in cracks and crevices in the rock and reduce water leakage into the tunnel. An overburden of 1 to 3 m of loose sediments appears to be well suited to stop leakage.

Large amounts of loose rock and sediments can however create a problem:
– The break-through takes place in an underwater scree or heap of stones and the tunnel opening is blocked by boulders, or large stones get stuck at the gate.
– The break-through takes place in a thick layer of sediments and loose stone masses which partly fills the tunnel.

46

Figure 9.1. Subsea tun-
nelling – lake taps/tunnel
piercings

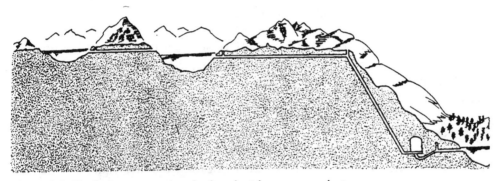

Figure 9.2. Typical cross-section, hydro electric power project.

The planning of a tunnel with break-through under water requires careful inspection in all phases of planning.

The first phase will always be a geological survey. A study by arial photogra-
phy will give information on rock structure, cracks and fissures in addition to information about the shape of the landscape and the location of loose rock and
sediments. This information together with a general survey of the area will provide background data for alternative break-throughs and serve as a basis for further investigations.

Uncomplicated break-throughs may only require an evaluation of the landscape, the geology and a ground survey.

A more detailed investigation will require a seismic survey. Profiles produced
by seismic refraction are to be preferred. This method differentiates between
zones of different acoustic velocity and will accurately define boundaries between
loose masses and solid rock.

The acoustic velocity of the rock will also give information about the rock quality.

Visual inspections can be carried out by divers or by remotely controlled submarines. Diving is used mostly in shallow waters. Remotely controlled submarines are to be recommended for depths of 50 m and more.

9.3   TUNNEL DRIVING TOWARDS THE BREAK-THROUGH, TEST DRILLING

The driving of a subsea tunnel towards the break-through is done with conventional drilling and blasting techniques. Extensive test drilling and grouting are usually required. Only conventional drilling and blasting methods will provide the necessary flexibility and give enough space at the tunnel face for these operations.

The tunnel is normally driven within 30 to 100 m of the break-through, partly dependent on the local conditions. The roof thickness above the tunnel should be at least twice the tunnel height. Test drilling should be used to check this at regular intervals.

Reduce the strain on the rock as much as possible by careful blasting. The rounds are gradually reduced to 2-1.5 m. The face is blasted in subsequent steps by slashing from a small pilot tunnel which is driven first.

Test drilling (Fig. 9.3) should be done on a regular basis within the last 30 to 100 m of the plug. It is normal to test drill with 4 to 5 holes in every second round. The test holes should be drilled to a depth of at least twice the length of the round so that the drill holes will overlap each other, alternatively the test holes should be long enough to cover a rock zone ahead of the face equal to twice the tunnel height or diameter.

The object of the test drilling is primarily to locate:
– Weak zones,
– Cracks and pockets with running water so that the necessary precautions can be taken such as rock bolting and grouting,
– To increase safety of the operation by checking the thickness of the roof above the tunnel and in front of the face.

Generally, at least one test hole should be drilled for each round when not under water and 4-5 holes for each round in the sub-sea part of the tunnel. The extent of drilling is increased in zones of poor rock quality or excessive water seepage. Procedures to measure water leakage, amount of clay in the flushing water and the reporting of abnormal conditions must be worked out beforehand to get the most out of the test drilling. The workforce at the tunnel face must be instructed to stop work promptly when an indication of excessive zones of poor rock quality/clay or water seepage occurs. A break-through into such zones of clay or heavy water seepage can have catastrophic consequences.

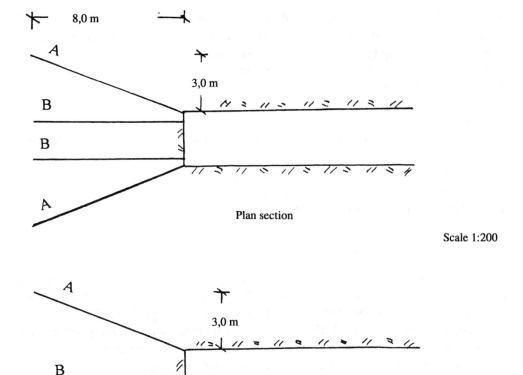

Plan section

Scale 1:200

Longitudinal section

Figure 9.3. Exploratory drilling procedure.

Conventional test drilling with drill rig registers primarily water seepage. Zones with crushed rock and clay can, to a certain extent, be registered by the rate of advance of the drill steel and the flushing water. The observations are less accurate when the holes are deep.

The most dependable information about rock quality, cracks and zones of clay is obtained by core drilling. Core drilling, although time consuming, gives good information and control.

9.4 METHODS FOR IMPROVING ROCK STABILITY

9.4.1 Rock bolting

The use of rock bolting is the most effective and commonly used technique to im-

prove stability. The technique is used in formations with cracks and fissions, and in rock of schistose structure. Rock bolting is used in the following two ways:
1. Systematic bolting to stabilize the rock surface,
2. Rock bolting in localized areas to stabilize blocks and fissions.
Three types of rock bolts are available:
1. Pre-stressed bolts, secured by expansion or with polyester. These bolts are effective immediately,
2. Grouted bolts. They require a time period for hardening and are for permanent installations.
3. Tubular bolts which are grouted when installed or later.
In badly cracked rock, the bolting is supplemented by straps and wire netting to secure the rock mass in between the bolts.

9.4.2   *Shotcrete and cast concrete lining*

The use of shotcrete or sprayed on concrete is an effective method to stabilize badly cracked structure and weak zones. The method is primarily used as a temorary measure to allow permanent stabilization to be performed later in the schedule. Shotcrete will gradually be dissolved in zones with running water, specially in salty water. Consequently, shotcrete is not a permanent measure.

The alternative will in most cases be a cast concrete lining.

All concrete is broken down by running water, especially by sea water. Water must be drained off before concrete is sprayed on. The same applies to cast concrete linings.

Shotcrete is an effective way of stabilizing zones containing clay deposits.

The pouring of a complete concrete lining can be the only solution under extreme conditions, often in combination with other means of stabilizing the rock such as rock bolting and shotcrete. In zones with running water, there is a risk that the concrete lining may cause the water pressure to build to an unacceptable level. In such cases, the water must be piped out through pipes running through the scaffolding. The piping is fitted with valves which can be closed for subsequent grouting work.

9.4.3   *Water protection*

Grouting is considered to be the only safe method to stop seepage of water through the rock structure. Grouting will have a stabilizing effect in most cases.

There are two types of grouting material available:
1. The concrete based grout which is the most commonly used. It is well suited to fill open cracks.
2. The chemical grouts which have better intrusion and hardening properties, but at higher cost which limits their use.
The extent of grouting is decided in each case as work proceeds.

Extensive water leakage combined with poor stability of the rock structure may require the casting of a double concrete lining. The design consists of two concrete linings with a watertight membrane of plastic or asphalt impregnated material between them.

9.5 METHODS OF BREAK-THROUGH

The decision as to the method of break-through should be taken during the planning stage. It may be necessary to make an adjustment of the tunnel route and its geometry to secure a successful result.

A differentiation is made between the two main methods of break-through, i.e. the open and the closed system.

The procedure for preliminary investigations, test drilling, blasting etc. is the same for the two systems.

With the open system there is direct connection between the tunnel face at the plug and the atmosphere through the gate shaft. The tunnel is closed either by a gate or by a concrete plug positioned downstream of the gate shaft to prevent an uncontrolled flow of water or rock masses through the tunnel.

The concrete plug is removed by blasting after the break-through has been made. The principle of an open system is shown in Figures 9.4 and 9.5, either with a closed main gate or with a temporary concrete plug downstream of the gate shaft, respectively, both in a water-filled tunnel and with air pocket below the break-through plug.

Advantages and disadvantages of the *open system* can be summarized as follows:

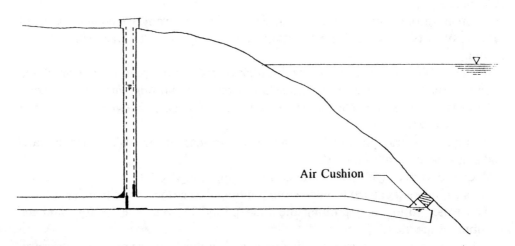

Figure 9.4. Open system of submerged tunnel piercing.

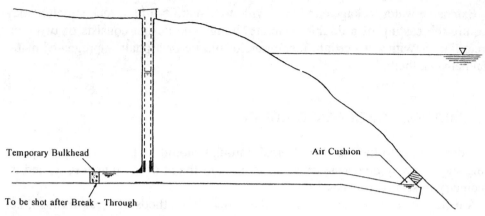

Figure 9.5. Open system of submerged tunnel piercing.

Advantages:
– The hydrodynamic conditions are clearly set out.
– Controllable concentration of the rock masses from the break-through plug and sediments above, if any.
– The pressure rise against the gate can be calculated to a fair degree of accuracy.

Disadvantages:
– Complicated procedure/arrangement for the filling of water, air, measurements etc.
– Considerable time lag between loading and shooting of the break-through plug.
The principle of the closed system is illustrated in Figures 9.6 and 9.7, with the shooting of the plug into a partly water filled tunnel or a dry tunnel, both with the gate closed.

A break-through into a dry tunnel, Figures 9.7, is from a technical point of view easy to perform, and the shooting of the plug can take place shortly after loading.

Water of high velocity will carry the rock masses of the plug uncontrolled and deep into the tunnel unless precautions are taken to avoid this. Further, unacceptable pressure rise at the gate can be generated if the distance between the plug and the gate is short.

The gate construction can be protected by various means. Well packed snow and straw have even been used.

A closed system with partly water-filled tunnel, Figure 9.6, will keep the rock masses from being carried too deep into the tunnel as compared with a dry tunnel.

The water filling will prolong the time period between the loading and the shooting of the plug.

Advantages and disadvantages of the *closed system* – dry tunnel – can be summarized as follows:

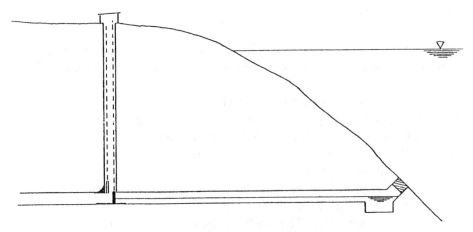

Figure 9.6. Closed system of submerged tunnel piercing.

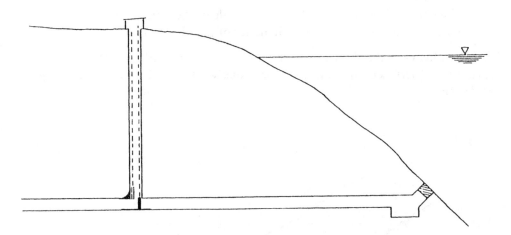

Figure 9.7. Closed system of submerged tunnel piercing.

Advantages:
– Uncomplicated design.
– Short period of time between the loading and the shooting of the plug.

Disadvantages:
– Uncertain hydrodynamic conditions.
– The rock masses from the plug and sediments above, if any, are carried uncon-
  trolled deep into the tunnel, and can cause damage to the gate construction.
– Requires considerable distance between the plug and the gate.

9.6   THE BREAK-THROUGH PLUG

The thickness of the plug is between 1 m in solid, dry rock and shallow water, and more than 10 m in poor quality rock and large cross-section with seepage of water.

The deposit above may vary from bare rock to a thickness of 5 to 6 m. Loose deposits of thickness 1-5 m appears to be ideal. The deposit will fill into cracks and fissures and reduce water leakage problems (Fig. 9.8).

The point of break-through should be located in a zone of good rock quality with moderate amount of deposits above.

The plug should be preferably rectangular in shape with rounded off corners to ease the break-through. The rectangular shape will simplify systematic drilling and positioning of the boreholes. The work will be easy to supervise and control.

The thickness of the tap can under normal conditions, with good quality rock and water depth in the region of 10-80 meters, be determined by the following rule of thumb:

Thickness of the plug = 1.2 × shortest side of the plug (rectangular)
Thickness of the plug = 1.0 × diameter of the plug (circular)

Figure 9.9 illustrates the thickness of the plug as a function of its cross-section.

Based on experience and data from previous work, the following curves have been drawn:

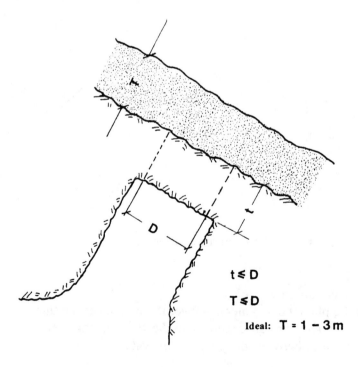

$t \leqslant D$

$T \leqslant D$

Ideal: $T = 1 - 3 \, m$

Figure 9.8. Break-through of the final plug.

- Specific loading as a function of the cross-section of the plug (Fig. 9.10),
- Specific drilling as a function of the cross-section of the plug (Fig. 9.11),
- Required area for large diameter holes as a function of the cross-section of the plug (Fig. 9.12).

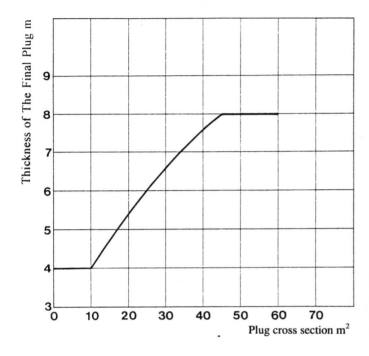

Figure 9.9. Thickness of the final plug as a function of the cross-section of the plug.

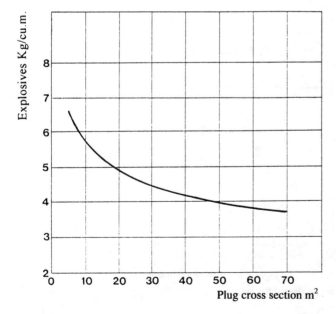

Figure 9.10. Specific loading.

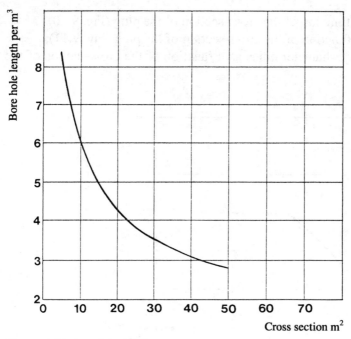

Figure 9.11. Specific drilling.

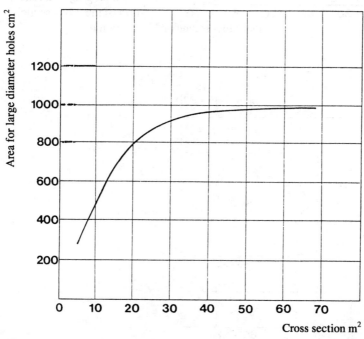

Figure 9.12. Required area for large diameter holes.

9.7 DRILLING OF THE BREAK-THROUGH PLUG

For a good result it is essential that the plug is closely examined by accurate test drilling. Equipment to plug holes which are drilled into the water must be readily available and tested beforehand.

The break-through should be made as close to a right angle to the seabed as possible. This will simplify the drilling of the plug.

The level of the tunnel should be positioned at a level so that the break-through gives a shaft of limited length, 5-10 m, which will provide sufficient air cushioning under the plug when the tunnel fills with water (Fig. 9.13).

The borehole diameters of the loaded holes are 45-64 mm. The large diameter holes (unloaded) have a diameter of 102 mm.

The drilling should be done through a template to achieve precision and avoid contact between adjacent holes.

All holes are drilled within 0.5 m of solid rock of the storage lake. It is a precondition that the contour of the rock surface and the length of the plug across the entire cross-section are known (Fig. 9.14).

The cut is placed in a position where it is expected that the quality of the rock is at its best and the hole is the least. The hole is the least where the length of the plug is at a minimum.

The curves of Figures 9.10, 9.11 and 9.12 are based on experience and should be used as bases for the drilling and loading plan.

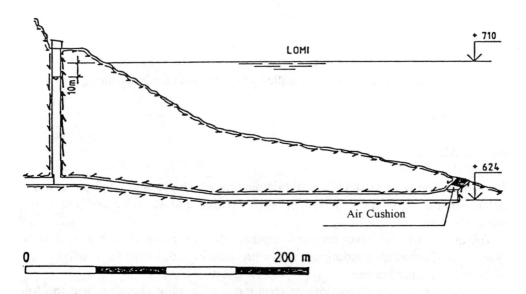

Figure 9.13. Section of head race tunnel.

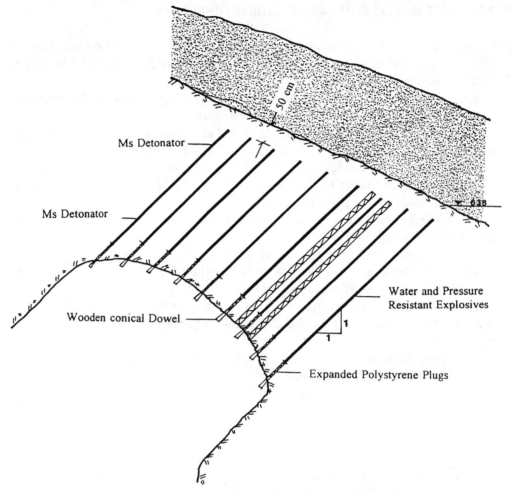

Figure 9.14. Final plug, the holes to be drilled within 50 cm of the break-through.

9.8 LOADING

For plug shooting use explosives and detonators which are water and pressure resistant for the actual time period between loading and shooting and the actual water pressure.

The quality of explosives required depends upon the cross-section, plug length, type of rock, amount of sediments, water pressure in addition to the configuration of the plug and the burden.

To find the quality of explosives required for the plug shooting, use the following rule of thumb: Twice the normal use in the tunnel up to the plug, plus 0.01 kg/m³ in addition for each meter of head of water pressure. Additional quantities in case of large amounts of deposits must be considered. Finally, an addition of 10% for plugged/lost holes.

Two detonators of the same number in each hole. It is recommended to use specially designed water and pressure resistant electric millisecond detonators.

9.9 EXPLOSIVES – DETONATORS

The manufacturer should test the explosives and the detonators beforehand to make sure that they fulfill the specifications.

Explosives and detonators are ordered and supplied with reference to and suitable for the actual pressure and time period under water.

The actual explosives to undergo functional tests:
– Crushing of lead block (Hess' test) before storage under water,
– Crushing of lead block (Hess' test) after storage under water at the actual depth and length of time,
– Detonating velocity after storage under water at the actual depth and length of time,
– Gap/sensitivity tests after storage under water at the actual depth and length of time.
The detonators to be tested in the following way. Two detonators of each interval ordered are tested for insulation resistance after storage under water at the actual depth and length of time. None of the detonators must show less than 30 M ohm. If none of the detonators fail in the test shooting, the consignment is approved for the plug shooting in question.

9.10 STEMMING

Polystyrene plugs, 5-10 cm long are used as stemming in the unloaded parts of the borehole. Solid wooden plugs with a notch for the detonator leads are used to keep the explosives and the stemming in place.

9.11 HOOK-UP DETONATORS

The hook-up of the detonators must be done accurately according to plan and in equal series which are coupled in parallel. All joints to be insulated in a watertight manner. The series shall be tested with approved ohmmeter. The resistance of each series shall be as equal as possible. The deviation shall not be more than ± 5%.

9.12 SHOT FIRING CABLES

The shot firing cable must be new and of high quality, and of the right dimension for the shot and the voltage of the exploder.

The place of firing is usually at the top of the gate shaft.

The shot firing cable should have a dimension of at least 2×2.5 m^2. The joining of cables should be done carefully and with a distance between the joints of at least 2 m. The joints to be insulated and made water proof by the use of crimping sleeves with silicone and tape.

It is assumed that the exploder used will have a low exit voltage and be of sufficiently high capacity.

9.13 PRESSURE – FLUSH OF STONE MASSES

9.13.1 *Open system*

This system requires a certain level of water to reduce the flush of stone masses towards the gate, and to reduce the upsurge into the gate shaft following the plug shot.

The degree of waterfilling will have a decisive effect on the design pressure. With a high degree of waterfilling the entire tunnel will be full of water with the exception of the established air pocket below the tap.

The design pressure, i.e. the explosion gas pressure, is the pressure which is generated when the gas of the explosion expands into the air pocket. This gas pressure would without the air pocket be propagated more or less undamped through the waterfilled tunnel to the gate, and most likely cause great damage to the gate structure.

9.13.2 *Closed system*

A high degree of water filling will give the same design pressure as in the open system.

Low water filling means that the tunnel is less than half full of water, 20% full is normally used. In this case the design pressure is generated by the water flowing through the tap opening and compressing the air in the tunnel. The increasing pressure will decrease the in-flow of water.

The pressure will reach a maximum when the in-flow of water comes to a stop. At this point the pressure of the entrapped air is higher than the outside water pressure, and the water starts flowing out of the tunnel.

9.13.3 *Explosion gas pressure*

The explosives in the plug will on detonation create a gas pressure in the air pocket which is dependent on the total quantity, pressure and volume of the air pocket. 1 kg of explosives will when detonated give approx. 0.8 Nm^3 of gas (1 Nm^3 is the quantity of gas which will give a volume of 1 m^3 at 1 atmosphere of pressure and at 0°C).

The explosion gas will expand into the air pocket and the pressure in the pocket will increase rapidly. The pressure rise will propagate through the water towards the closing device (gate, concrete plug) and be reflected. The rise of pressure at the closing device will be twice the amplitude of the pressure wave.

For design calculations the following is assumed:
– All of the explosion gas from the shooting of the plug will expand into the air pocket. This takes place so rapidly that the volume of the air pocket can be assumed constant during detonation.
– No damping effect of the pressure wave takes place on its way to the closing device.

Symbols used in pressure calculations:
– *Pi* stands for the pressure in the air pocket prior to shooting,
– *Pg* stands for the peak pressure immediately after the shooting,
– *Pl* stands for the static pressure against the closing device prior to shooting,
– *Pd* stands for the design pressure against the closing device.

The pressure rise in the air pocket immediately after shooting:

$$Pg - Pi$$

When the shock wave hits the closing device and is reflected, the pressure will rise by:

$$2 \times (Pg - Pi)$$

Design pressure against the closing device:

$$Pd = Pl + 2 \times (Pg - Pi)$$

The amount of air below the plug is controlled during the water filling operation.

A surge of water in the gate shaft will be generated if the shaft is filled up to a level higher than the water level in the water magazine.

The upward surge in the shaft can reach 70-90% of the difference between the level of the magazine and the level to which the gate shaft has been filled, depending on frictional losses.

Refer to Figure 9.15, Open system – surge in gate shaft, and Figure 9.16, Gas pressure in air pocket.

9.13.4 *Example – calculation of explosion gas pressure – open system*

The calculations are based on the drawing in Figure 9.17.

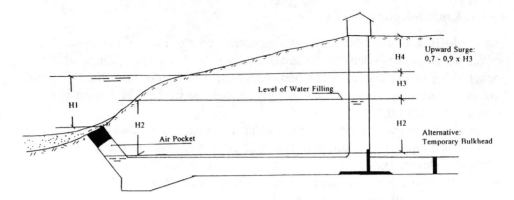

1. To Avoid Blow-out of The Air Pocket: H2 $<$ H1

2. To Avoid An Up-Surge into The Gate House: H4 $>$ c x H3 (c Normally 0,7 - 0,9)

Figure 9.15. Open system, surge in gate shaft.

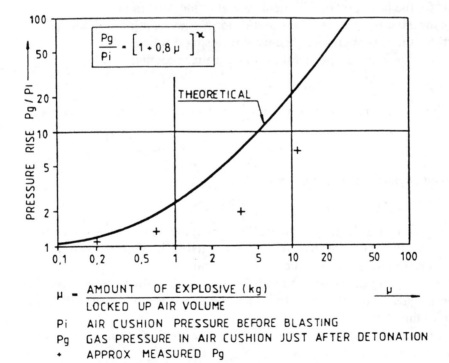

Figure 9.16. Gas pressure in air pocket (open piercing), after Solvik 1981.

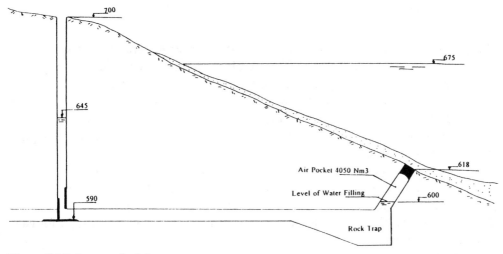

Figure 9.17. Layout for lake tap.

The assumptions for the pressure calculations are:

Water level, storage lake	Contour line 675
Water level, air pocket	Contour line 600
Water level, shaft	Contour line 645
Gate sill	Contour line 590
Entrapped air at the tunnel face	4050 Nm^3

Total quantity of explosives in the plug shooting: 1350 kg

Pressure against gate prior to shooting: $Pl = (645-590)/10 + 1 = 6.5$ bar abs
Pressure in air pocket prior to shooting: $Pi = (645-600)/10 + 1 = 5.5$ bar abs
u = quantity of explosives in kg/entrapped air Nm^3 = 1350 kg/4050 Nm^3 = 0.33

Refer to Figure 9.15. The ratio between quantity of explosives and quantity of entrapped air at the face is decisive for the resulting explosion gas pressure. The quantity of explosives is usually fixed beforehand, while the quantity of entrapped air must be adjusted to derive at an acceptable pressure rise.

$$Pg/Pi = (1 + 0.8\ u) = (1 + 0.8 \times 0.33) = 1.4$$
$$Pg = Pi\ /\times 1.4 = 5.5 \times 1.4 = 7.7 \text{ bar}$$

Pressure rise in the air pocket on shooting: $Pg - Pi = 7.7 - 5.5 = 2.2$ bar

Design pressure against gate: $Pd = Pl + 2\ (Pg - Pi) = 6.5 + 2 \times 2.2 = 10.9$ bar

This corresponds to the pressure of a water column of 109 m above the gate sill.

9.14   CLOSING REMARKS

Plug shooting under water is always a most exciting undertaking. It is the final step of the operation and draws considerable attention. A number of important details must however be in place before the shooting can take place.

The driving of a tunnel under water and the preparation for a break-through with plug thickness of 4 to 8 m are difficult operations. The work requires careful preliminary inspection and deliberation, and accuracy in every detail of the planning and shooting procecss.

Superficial inspection, poor planning and workmanship can give rise to dramatic and costly surprises. Uncontrolled water leakage filling the tunnel or an uncontrolled landslide of loose masses blocking the break-through exit is normally the result of insufficient preliminary inspection of poor workmanship, with catastrophic consequences.

Nevertheless, deviation from the original plan may be required on the way regardless of the accuracy of the planning process. In such cases it can be of decisive importance to have experienced people on the spot so that the new situation can be evaluated and appropriate corrective action taken to avoid critical situations.

Preparation for the break-through shooting, loading, water filling etc. is a busy and hectic period and the work must proceed continually round the clock. There may be considerable water seepage, and it is a wet, cold and unpleasant job to do at the plug face. Willingness to work and endurance are essential qualities for everyone engaged in this operation.

It is important that the entire crew is well prepared and made familiar with the various operations and safety procedures. Procedures for loading, hook-up and air-filling etc. must be prepared beforehand and fully discussed in advance. This is followed up by the use of control points along the road which are checked off by signature.

CHAPTER 10

Tunnel boring machines

In Norway tunnel driving with tunnel boring machines, TBMs, started some 25 years ago. Ten years later approximately 10% of the tunnels below 30 m^2 in cross-section were driven with TBMs. The peak was reached in 1983 with 30 km of TBM driven tunnels that year. In subsequent years the use of TBM was gradually reduced in favour of conventional methods.

Today, the TBMs are primarily used for smaller shafts for cables and ventilation, and for water tunnels with cross-section area between 2 and 30 m^2.

The TBM leaves a tunnel with a circular cross-section which is not the best for road tunnels. Enlargement of the lower part of the tunnel by blasting is necessary to give required space for the height and the width of the road. Nevertheless, the TBMs produce an even contour which is valuable for road tunnels. By experience, however, the enlargement of the lower part of the tunnel requires blasting well beyond the theoretical profile so that part of the even contour left by the TBM is lost. So far, only one road tunnel has been driven in Norway using TBM and that was 10 years ago.

The TBM is vulnerable to unexpected circumstances, caused either by poor planning, insufficient preparatory work or unforeseen rock structure.

To avoid time consuming delays and hazardous working conditions, it is important to keep a close eye on the rock conditions during the TBM operation. The same procedure for inspection and testing used during conventional tunnel driving must be followed (Fig. 10.1).

Figure 10.1. To prevent hazardous situations – keep a close look and inspect the rock condition at regular intervals. This also applies when using TBM equipment.

CHAPTER 11

Controlled and cautious blasting

11.1   ACCURACY OF DRILLING

Good overall financial result of a blasting job, including safe performance and high quality work, depends to a large extent on the accuracy of the drilling.

An accurately drilled blast will save borehole meters and consumption of explosives, and give optimal results with respect to fragmentation, muckpile and vibrations. In addition it will give correct level of the floor and an even contour surface with a minimum of back breakage of the remaining rock.

11.2   CONTOUR

In tunnelling, the contour surfaces are the permanent walls and roof along and below which must be safe to walk.

The object of the contour blasting is to preserve the theoretical profile and reduce subsequent scaling and support of the contour surfaces.

In principle, there are two methods for contour blasting: Smooth blasting and pre-split blasting. Either method is based on the drilling of holes along the contour surface with fairly short borehole distances and with the use of reduced explosives loading. Contour blasting leaves even surfaces when the borehole distances, type of explosives and explosives loadings are carefully adjusted to the geological conditions and provided that the boreholes are accurately drilled and in parallel.

Smooth blasting is mostly used partly because the result is more predictable and partly because the vibration set up is less as compared with pre-split blasting.

11.3   CONTROLLED BLASTING

Controlled blasting is necessary to preserve the stability of the remaining rock and attain a minimum of scaling and rock stabilization. The area of damage and the amount of back-breakage are directly related to the type and concentration of explosives in the boreholes of the contour. By experience, ANFO in the contour holes will give more damage than with the use of tube charges. Best results are obtained when both the contour and the first row of holes are loaded with tube charges.

The following should be observed to keep vibration at a low level and maintain good quality of the remaining rock.

– All boreholes to be drilled accurately to the predetermined depth and direction (Fig. 4.5).

– The cut is drilled with at least 4 large diameter holes to ease the breakage for the loaded holes (Fig. 4.6).

– The contour holes and the lifter holes which are drilled at an off-set angle must be given an easy break. This is done by drilling the first row on the inside of the contour and the lifter holes in parallel. The contour holes (the outer row) are drilled 60 cm shorter than the rest of the holes (Fig. 4.1).

The amount of loading per detonator interval is, in addition to the accuracy of drilling, very important for the level of vibration (Figs 4.11, 4.12, 5.1, 5.2 and 5.3).

It is important in cautious tunnel blasting that the interval time of the detonators are long enough for the rock to be properly broken and blasted away from the face before the next hole is fired. Nonel LP is specially developed for tunnel blasting with suitable time intervals and interval numbers 0-60. See Figure 8.3.

For bench blasting in rock caverns with the Nonel Unidet, the number of interval possibilities represents a good initiation system to keep the vibrations under control.

During the portal blasting the tunnel will have limited overburden. Cautious blasting is required with reduced length of the round and with blasting of the cross-section in two stages with auxiliary tunnel and back ripping (see Fig. 2.1). Cartridged explosives are required in this phase of the work to keep the loading per interval at an acceptable level.

11.4   BLAST VIBRATION IN GENERAL

The strain on a structure caused by vibration from a blast is in general dependent on the following three factors:
1. The sequence of the detonation,
2. The properties of the material transmitting the energy from the blast to the structure, and
3. The type of foundation.

When a blast breaks rock against a free face, most of the energy liberated is used to break loose the rock. Part of the energy will, however, be used to set up transitory vibrations in the adjacent rock. The amount of the liberated energy which sets up vibrations will depend on the degree of constraint of the loading. An easy break (low burden) will give less vibration than the same charge in a heavy break (high burden).

The energy liberated in a blast which inter alia sets up vibrations is dependent on the type of the blast and its size, borehole dimension, drilling pattern, specific loading and sequence of initiation.

Nothing can be done with the properties of the transmitting material. To which extent the energy of the vibration waves will reach the structure in question depends on the physical properties of the rock and possible loose masses in between (type of rock, amount of bedding, cracks, grain distribution, ground water level etc.). An accurate evaluation of these factors is difficult.

A numerical calculation of the actual strain on a given structure must to a great degree be based on past experience, assumptions made and empirical equations. Experience from past and comparable work and possible prior trial blasts will be of great value in the evaluation.

Vibrations set up by a blast will propagate through the medium with the local acoustic velocity and concentrically in all directions. The intensity will diminish with the distance from the point of initiation. Surface vibrations are generated when the wave reaches the surface. A surface particle will move in an elliptical path with the major axis in the vertical plane.

Irregularities in the rock structure make the course taken by the wave very complicated.

Definitions

Vibrations:	Periodic motion in alternate directions and propagation from point to point.
Amplitude-movement:	Particle movement – measured in microns. Measured as maximum movement from the base line.
Particle velocity:	Particle velocity (v) – measured in m/sec.
Acceleration:	Particle acceleration – measured in m/sec^2
Frequency:	Vibrations per unit of time measured in Hz – number of vibrations/sec.
Velocity of propagation:	Velocity of propatation of the shock wave – seismic velocity, acoustic velocity – measured in m/sec.

Limiting values based on damage criteria are derived from long time experience with the relationship between the vertical particle velocity and the effect on housing and installations under different ground conditions.

The vibrations from a blast reach very quickly maximum amplitude and is thereafter rapidly damped to a minimum within a few milliseconds.

Detonators with intervals between 50 and 500 milliseconds are normal for underground work. Consequently, it is the charge per interval which determine the amplitude.

The frequency of the vibration will be high, 500-1000 Hz, a few meters away from the charge. Further away, 10-15 m, the same shot will give vibrations of frequency between 10 and 100 Hz.

The frequency falls rapidly with distance.

The movement of a discrete particle which takes part in the vibration can be described with the three parameters: The amplitude of the particle from the baseline, frequency of the particle vibration and the particle acceleration.

The vibration at a point at a distance from the charge is proportional to the weight of the charge and inversely proportional to the distance.

The rock condition and the type of rock between the charge and the point of measurement effect the amount of vibration. Such factors are bedding, strike and dip, fissures and zones of weakness.

11.5   LEVEL OF VIBRATION AND CONTROL

The following general empirical equations gives the level of vibration in terms of particle velocity:

$$v = K \times (Q^\alpha / d^\beta)$$

Where v = particle velocity in mm/s, K = rock constant, Q = charge in kg per detonator delay number, d = distance in m from the center of gravity of the blast to the point of measurement = coefficients of damping.

The coefficients of damping is difficult to evaluate. The equation gives for this reason unsatisfactory results. For practical purposes a simpler equation will give better results:

$$v = K \times \sqrt{Q} / d$$

with K as the only variable coefficient. The rock constant can for example vary between 50 (long distance) and 500 (short distance).

The blast vibration at a point away from the shot is proportional to the weight of the charge fired simultaneously and inversely porportional to the distance. The frequency falls rapidly with distance. Conditions like the burden, inaccuracy of the drilling with respect to the direction and depth, relationship between the charge and the specific loading are factors of importance for the level of vibration.

The use of vibration recorders is necessary to achieve optimum results within the limits given to the construction company. The vibration recorder shall register the entire sequence of the detonation and record the parameters as described. This makes it possible to analyse the sequence of detonation and find out at which point in the blast the vibrations reach a maximum so that the shooting plan can be promptly adjusted.

A heavy burden caused by inaccurate drilling of cracks filled with ANFO will give increased level of vibration.

When the distance between the blast and the point of measurement is less than 5 m no relationship exists between the level of vibration and the risk of damage on existing buildings. In such cases it is the movement and lift of the rock which are the decisive factors.

When the shooting plans are being worked out and during the initial phase of the work, use rock constants based on experience or apply blast design data from a previous job which is comparable with the present project.

Appendices

Principle sketch of underground chambers

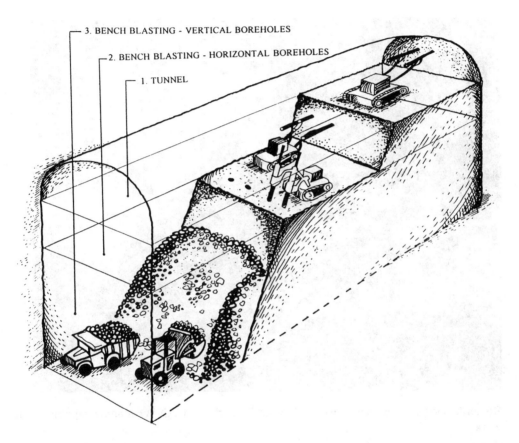

3. BENCH BLASTING - VERTICAL BOREHOLES

2. BENCH BLASTING - HORIZONTAL BOREHOLES

1. TUNNEL

Blasting underground chambers.

Drilling and loading (anfo) equipment

Romeriksporten – railway tunnel. AMV drilling rig and ANFO loading equipment at work.

APPENDIX 3

Loading with ANFO and NONEL detonators

NONEL detonators and primers are placed at the bottom of the boreholes. Loading the boreholes with ANFO explosives is in progress.

Equipment for pumpable emulsion explosives in tunnels (site mixed slurry)

Loading truck for bulk emulsion explosives.

Loading truck for bulk emulsion explosives.

APPENDIX 5

Plug blasting – the Lomi Hydro Electric Power Plant

A5.1   GENERAL DESCRIPTION

Lomi Hydro Electric Plant is located in the northern part of Norway, approximately 100 km east of Bodø.

The headrace tunnel was connected to the lake Lomi by the 'Lake-Tap' method.

The static head of water on the plug was 75 m. The cross section of the final plug was 18 m^2 and the thickness of the plug about 4.5 m average. The plug was covered by 2-3 m of loose deposits. The distance between the plug and the gate shaft was 280 m (Fig. A5.1). The rock type was quartzitic mica schist. The leakage of water through the rock close to the final plug was insignificant.

At the time of blasting the gate was closed, and the tunnel system was filled with water to 10 m below the level of the lake. Below the plug an air cushion was established.

A high strength gelatine explosive and millisecond delay detonators were used for the final blast.

A5.2   DRILLING PATTERN

The drilling pattern was based on the use of parallel holes with 3 separate parallel hole cuts.

In each of the parallel hole cuts there were 4 uncharged 5 inch diameter large holes, the rest were 45 mm holes (Fig. A5.2).

Number of boreholes: 80-45 mm, charged; 12-5 inch, uncharged.

The plug thickness was 3.9.5.0 m.

The boreholes were drilled 0.5 m short of break-through of the plug, i.e. an average length of the boreholes of 4.0 m.

A5.3   CHARGING AND IGNITION SYSTEM

The boreholes were loaded with 35 × 600 mm cartridges in plastic film, contain-

80

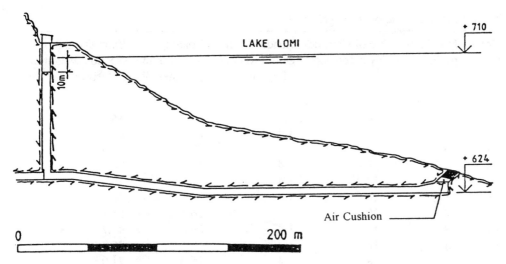

Figure A5.1. Section of headrace tunnel.

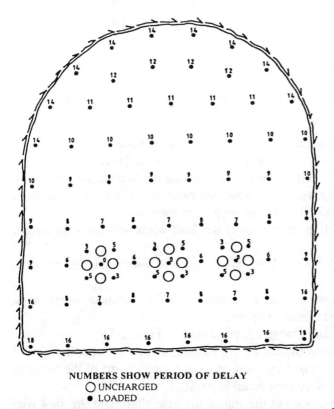

NUMBERS SHOW PERIOD OF DELAY
○ UNCHARGED
● LOADED

Figure A5.2. Cross section, final plug.

Table A5.1. Delay detonators – charges.

Delay no	No of holes per interval	No of short delay detonators	Loading concentration kg/m of borehole	Loading length in m per interval	'Extra Dynamit' 35 × 600 weight of explosives kg/interval
0	3	6	1.4	10.3	14.4
3	6	12	1.4	20.6	28.8
5	6	12	1.4	20.6	28.8
6	4	8	1.4	14.5	20.3
7	6	12	1.4	21.0	29.4
8	8	16	1.4	28.8	40.3
9	10	20	1.4	36.5	51.1
10	8	16	1.4	29.8	41.7
11	5	10	1.4	18.4	25.7
12	4	8	1.4	15.0	21.0
14	10	20	1.4	37.5	52.5
16	8	16	1.4	28.1	39.3
18	2	4	1.4	7.4	10.4
Total	80	160			403.7

Total quantity of rock: 80 m^3
Total explosives required: 404 kg
Power factor 404/80: 5,1 kg/ m^3

ing 60% NG explosives, sold under the trade name 'Extra Dynamit'.

The charge concentration was calculated to be 1.4 kg per m borehole.

In all the loaded holes 2 millisecond delay detonators with identical delay number and normal sensitivity were used. One detonator was placed in the bottom of the hole, the other in the middle of the outer part of the hole.

The detonators had protective sheaths over the lead wires and the wire length was 6 m.

The unloaded part of the holes – 0.3 m – were stemmed with expanded polystyrene plugs.

A wooden conical dowel with a precut opening for the detonator wires were used to keep the charge and stemming in place.

Details of ignition system and charge are presented in Table A5.1.

The round was split into two circuits, i.e. 80 detonators in each circuit with the bottom detonators in one circuit and the detonators in the outer part of the holes in the other circuit. The circuits were connected in parallel.

The blasting machine was placed at the top of the gate shaft, and the shot was fired immediately after the water in the gate shaft reached a level 10 m below the lake level.

APPENDIX 6

Plug blasting – the Tyee Lake Hydro Electric Power Plant

A6.1 GENERAL DESCRIPTION

The Tyee Lake Hydroelectric Plant is located near the town of Wrangell in southern Alaska.

Break-through of final plug was carried out in September of 1983.

Static head on the plug was 50 m.

Distance between plug and gateshaft was 85 m.

Cross-section of plug was 8 m^2 and the plug thickness was 2.5 m to 3 m.

The rock type was granodiorite and extensive grouting was necessary to seal the plug area.

A6.2 DRILLING PATTERN

The drilling pattern was based on the use of parallel holes with 2 separate parallel hole cuts on the symmetric line through the center of the tunnel face.

In each of the parallel hole cuts there were 4 uncharged 76 mm diameter large holes, the rest were 35 mm diameter holes (Fig. A6.1).

The boreholes were drilled 0.3 m short of break-through of the plug, i.e. an average length of the boreholes of about 2.5 m.

A6.3 CHARGING AND IGNITION SYSTEM

The boreholes were loaded with 25 × 200 mm cartridges, containing 60% NG explosive, sold under the trade name 'Extra Dynamit'.

The charge concentration was calculated to be 1 kg/m borehole.

In all the loaded holes 2 millisecond delay detonators with identical number and normal sensitivity. One detonator was placed in the bottom of the hole, the others were used in the middle of the outer part of the hole.

The detonators had protective sheaths over the lead wires and the wire length was 4 m.

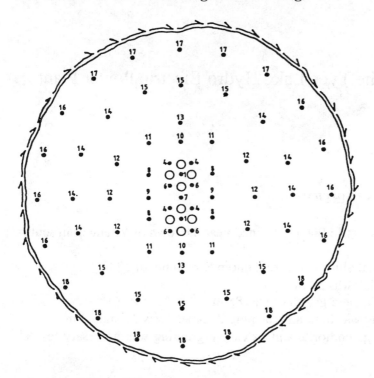

NUMBERS SHOW PERIOD OF DELAY
○ UNCHARGED
● LOADED

0 ═══════════════════ 3 m

Figure A6.1. Cross-section of final plug. Delay pattern using MS delays.

Table A6.1. Delay detonators – charges.

Delay no	Interval m/sec	No of holes per interval	No of short delay detonators per interval	Loading concentration kg/m of borehole	Loading length in m per interval	Explosives 25 × 200 mm kg/interval
1	25	2	4	1.0	4.4	4.4
4	100	4	8	1.0	8.8	8.8
6	150	4	8	1.0	8.4	8.4
7	175	1	2	1.0	2.2	2.2
8	200	4	8	1.0	8.8	8.8
9	225	2	4	1.0	4.4	4.4
10	150	2	4	1.0	4.4	4.4
11	275	4	8	1.0	9.0	9.0
12	300	6	12	1.0	13.2	13.2
13	325	2	4	1.0	4.4	4.4
14	350	8	16	1.0	17.6	17.6
15	375	8	16	1.0	17.0	17.0

Table A6.1. Continued.

Delay no	Interval m/sec	No of holes per interval	No of short delay detonators per interval	Loading concentration kg/m of borehole	Loading length in m per interval	Explosives 25 × 200 mm kg/interval
16	400	8	16	1.0	17.6	17.6
17	425	5	10	1.0	17.0	14.5
18	450	7	14	1.0	13.3	13.3
Total		67	134			143.6

Total quantity of rock: $1.6^2 \times \pi \times 2.8 = 22.5$ m^3
Total explosives required: 150 kg
Powder factor $143.6/22.5 = 6.38$ kg/m^3

A wooden conical dowel with a precut opening for the detonator wires was used to keep the charge in place.

Detailes of the ignition system and charge are presented in Table A6.1.

The round was split into two circuits, 67 detonators in each circuit, with the bottom detonators in one circuit and the detonators in the other part of the holes in the outer circuit.

Properties of the 'Extra Dynamit' explosives, results of underwater testing

Storing under pressure, in 7 days by 12-31°C.

		Water depth in m		
		0	100	200
Weight	Kg	5,68	5,68	5,67
Density	Kg/l	1,50		
Peak overpressure	Mpa	7,48	7,81	7,81
Shock energy	MJ/kg	0,88	0,86	0,84
Bubble energy	MJ/kg	2,02	1,84	1,86
Total energy	MJ/kg	4,63	4,38	4,35
Yield (100 × Total energy/theoretical energy)	%	87	82	82
Weight strength compared to LFB – Dynamite	%	93	89	88
Weight strength compared to dynamite	%	100	95	95
Weight strength compared to ANFO	%	115	110	109
Mean scatter between channel A and B for peak overpressure	%	2	2	0
Mean scatter between channel A and B for shock energy	%	5	7	5

APPENDIX 8

Waterproof connecting of detonator leads – 'Scotchlok' connectors

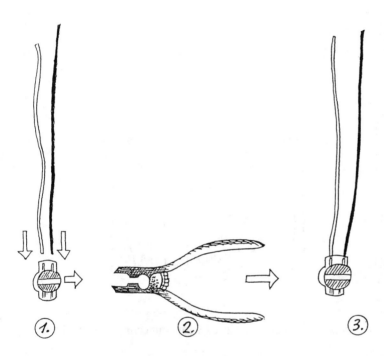

Procedure for waterproof connecting of detonator leads. 1. 'Scotchlok' connector (3M); 2. Pliers; 3. Waterproof connection.

APPENDIX 9

Waterproof connecting of detonator leads – connecting sleeves

CONNECTING SLEEVES

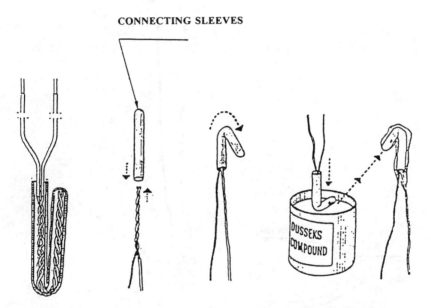

Procedure for waterproof connecting of detonator leads with connecting sleeves (Dyno).

Wooden conical dowel to keep the charge and stemming in place, suitable for 45-63 mm boreholes

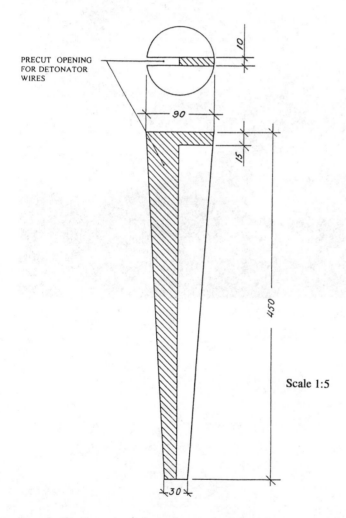

PRECUT OPENING
FOR DETONATOR
WIRES

90

10

15

450

Scale 1:5

30

Longitudinal section of wooden conical dowel.

Blasting final plug

Water column after plug blasting.

Lake Rembesdal after draw down – depth 25 m, cross-section 54 m^2, plug thickness, 2,5-7 m, soil cover 3-5 m

Opening after plug blasting and draw down, Lake Rembesdal (Norway).

APPENDIX 13

Kobbelv hydro power plant – blasting final plug

A13.1 FOSSVATN – FINISHED 1987

The Fossvatn lake is regulated between 10 and 90 m above and below normal water level. This gives a total height of regulation of 100 m.

The piercing into the Fossvatn lake is the deepest plug blasting of the Kobbelv project with a water pressure of 119 m at the tunnel gate. This is possibly the deepest plug blasting performed into a hydro electric power project up to now.

The drilling of the plug was done in accordance with a drilling plan worked out in advance. Only minor deviation from the original plan was necessary.

The rock structure in the break-through area was full of cracks and fissures. 15,000 kg of concrete was injected. The water leakage was, nevertheless, considerable. After drilling of the plug the leakage was in excess of 4.5 m³/min. This made the loading a rather wet undertaking. The drilling of the plug was completed in 1985.

A13.2 BLASTING PLAN FOR THE FINAL PLUG

The drilling and loading plan for the plug was worked out based on actual water depth, thickness of the plug, plug cross-section and amount of loose rock and sediments (Fig. A13.1).

The drilling pattern included two large diameter parallel cuts with a total of 8 large diameter holes. Otherwise, the pattern was set out as clearly and systematically as possible. The drilling pattern outside the cut was 60 × 60 cm.

All the holes were to be drilled within approximately 50 cm of break-through. This was done with the exception of a few holes where the depth was reduced leaving 5.5 m of rock because of a flaw with running water. This water leakage created certain problems later during the loading and hook-up process.

After completion of the drilling operation all bore holes were marked and recorded with respect to position and depth.

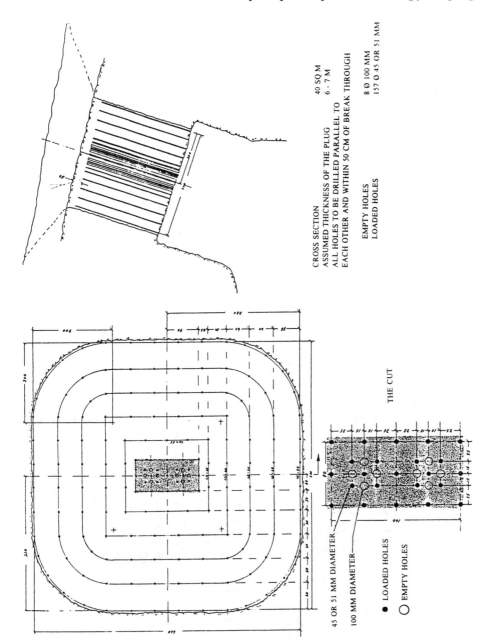

CROSS SECTION
ASSUMED THICKNESS OF THE PLUG 40 SQ M
ALL HOLES TO BE DRILLED PARALLEL TO 6 - 7 M
EACH OTHER AND WITHIN 50 CM OF BREAK THROUGH

EMPTY HOLES 8 Ø 100 MM
LOADED HOLES 157 Ø 45 OR 51 MM

THE CUT

45 OR 51 MM DIAMETER
100 MM DIAMETER

● LOADED HOLES
○ EMPTY HOLES

Figure A13.1. Kobbelv hydro power plant, blasting final plug, drilling pattern.

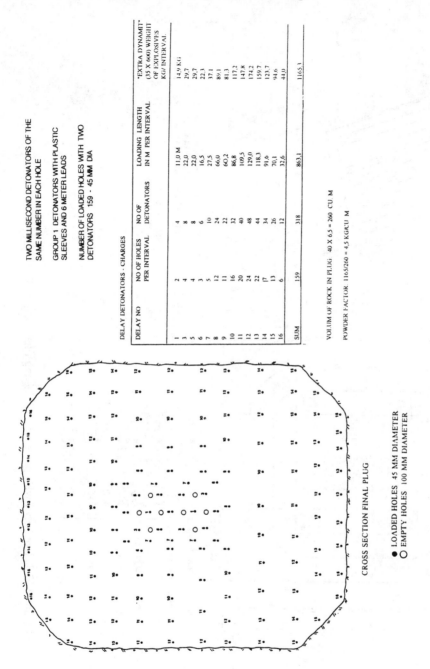

TWO MILLISECOND DETONATORS OF THE SAME NUMBER IN EACH HOLE

GROUP 1 DETONATORS WITH PLASTIC SLEEVES AND 6 METER LEADS

NUMBER OF LOADED HOLES WITH TWO DETONATORS 159 - 45 MM DIA

DELAY DETONATORS - CHARGES

DELAY NO	NO OF HOLES PER INTERVAL	NO OF DETONATORS	LOADING LENGTH IN M PER INTERVAL	EXTRA DYNAMIT (35 X 600) WEIGHT OF EXPLOSIVES KG/INTERVAL
1	2	4	11,0 M	14,9 KG
3	7	8	22,0	29,7
5	4	8	22,0	29,7
6	3	6	16,5	22,3
7	5	10	27,5	37,1
8	12	24	66,0	89,1
9	11	22	60,2	81,3
10	16	32	86,8	117,2
11	20	40	109,5	147,8
12	24	48	129,0	174,2
13	22	44	118,3	159,7
14	17	34	91,6	123,7
15	13	26	70,1	94,6
16	6	12	32,6	43,0
SUM	159	318	863,1	1165,3

VOLUM OF ROCK IN PLUG 40 X 6,5 = 260 CU M

POWDER FACTOR 1165/260 = 4,5 KG/CU M

CROSS SECTION FINAL PLUG

● LOADED HOLES 45 MM DIAMETER
○ EMPTY HOLES 100 MM DIAMETER

NUMBER AT THE BOREHOLES SHOW PERIOD OF DELAY

Figure A13.2. Kobbelv hydro power plant, blasting final plug – Fossvatn. Initiation plan – loading table.

A13.3 LOADING AND INITIATION PLAN, FIGURES A13.2-A13.5

The final loading plan was worked out after the plug drilling was completed and the bore holes marked and recorded.

Extra Dynamit in cartridge dimension 30 × 600 mm was used. The loading density worked out to be 1,4-1,5 kg/m. Approx. 50 cm of the holes were to be left to be unloaded.

Extra Dynamit was ordered specifying use at 100 m water pressure over a 72 hour period.

The total consumption of explosives was calculated to be 1200 kg. With 270 cu.m of rock this corresponds to a specific loading of 4.5 kg/ m^3.

The detonators used were of Group 1, German make, with lead wire length of 6 m and reinforced with plastic sleeves. The detonators were ordered for use at 100 m of water pressure over a 72 hour period.

Two millisecond detonators of the same number were used in all of the loaded holes.

A13.4 WORKING PROCEDURE FOR THE LOADING/HOOK-UP OF THE FOSSVATN PLUG BLASTING

Scaffolding: Wooden scaffolding was errected all the way up to the plug face. The height of the scaffolding was 18 m.

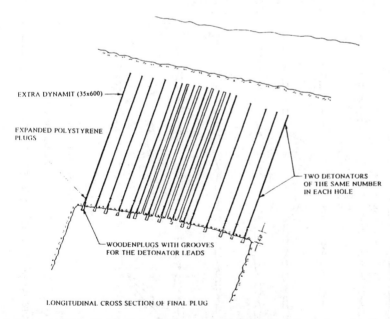

Figure A13.3. Kobbelv hydro power plant, blasting final plug, loading plan.

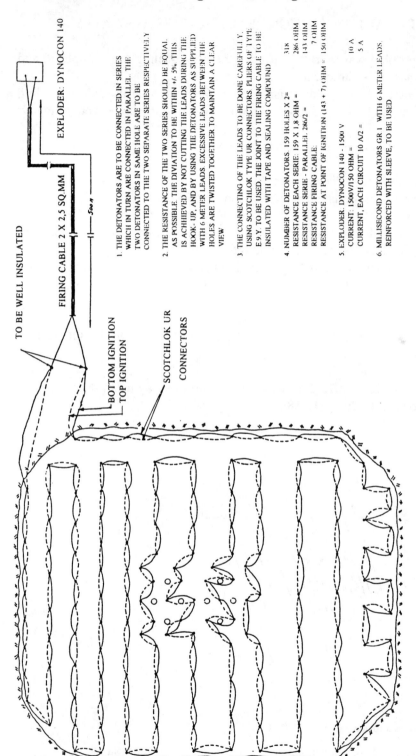

EXPLODER: DYNOCON 140

TO BE WELL INSULATED

FIRING CABLE 2 X 2,5 SQ.MM

500 m

BOTTOM IGNITION
TOP IGNITION

SCOTCHLOK UR CONNECTORS

1. THE DETONATORS ARE TO BE CONNECTED IN SERIES WHICH IN TURN ARE CONNECTED IN PARALLEL. THE TWO DETONATORS IN SAME HOLE ARE TO BE CONNECTED TO THE TWO SEPARATE SERIES RESPECTIVELY

2. THE RESISTANCE OF THE TWO SERIES SHOULD BE EQUAL AS POSSIBLE. THE DEVIATION TO BE WITHIN +/- 5%. THIS IS ACHIEVED BY NOT CUTTING THE LEADS DURING THE HOOK-UP, AND BY USING THE DETONATORS AS SUPPLIED WITH 6 METER LEADS. EXCESSIVE LEADS BETWEEN THE HOLES ARE TWISTED TOGETHER TO MAINTAIN A CLEAR VIEW

3. THE CONNECTING OF THE LEADS TO BE DONE CAREFULLY. USING SCOTCHLOK TYPE UR CONNECTORS PLIERS OF TYPE E 9 Y. TO BE USED. THE JOINT TO THE FIRING CABLE TO BE INSULATED WITH TAPE AND SEALING COMPOUND

4. NUMBER OF DETONATORS: 159 HOLES X 2= 318
   RESISTANCE EACH SERIE: 159 X 1,8 OHM = 286 OHM
   RESISTANCE SERIE - PARALLEL 286/2 = 143 OHM
   RESISTANCE FIRING CABLE: 7 OHM
   RESISTANCE AT POINT OF IGNITION (143 + 7) OHM = 150 OHM

5. EXPLODER: DYNOCON 140 - 1500 V
   CURRENT: 1500V/150 OHM = 10 A
   CURRENT, EACH CIRCUIT 10 A/2 = 5 A

6. MILLISECOND DETONATORS GR 1 WITH 6 METER LEADS REINFORCED WITH SLEEVE, TO BE USED

CROSS SECTION FINAL PLUG

CONNECTING DETONATORS

Figure A13.4. Kobbelv hydro power plant, blasting final plug, connecting detonators.

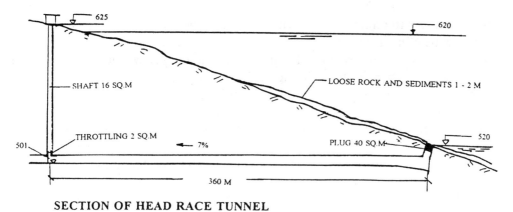

SECTION OF HEAD RACE TUNNEL

Figure A13.5. Final plug blasting Fossvatn, 03.04.87.

Loading: Each borehole was measured/controlled for correct depth. Correct number of cartridges was loaded in each hole. The detonators were normally placed in the second cartridge from the top and from the bottom. An isopor (expanded polystyrene) stopper was used as stemming in each hole. Statkraft (the contractor) wanted, if possible, to reduce the amount of explosives to be used and for this reason the unloaded part of the boreholes were somewhat increased.

The holes were plugged with wooden plugs one after the other after loading. The leading wires were fixed to the wooden plug.

Hook-up: The hook-up was done with Scotchlok hook-up clips, type UR.

The top and the bottom detonators were coupled to separate circuits. The circuits were kept apart from each other all the way to the gate house.

Firing cable: $2 \times 2,5$ mm^2. Three $2 \times 2,5$ mm^2 firing cables were run down the gate shaft. Two of the cables were stretched all the way to the plug. One cable was run some 100 m into the tunnel for the cutting of the shooting cables and the air supply hose.

Exploder: Cl 160 VA.

Water filling: Two pumps, 1000 m^3/hour each, were used. Filling of water from the Fossvatn lake via the shaft through two 30 m long 10" hoses.

Air filling: One 16 m^3 electric driven compressor; One 9 m^3 diesel driven compressor; Maximum air pressure equal 10 bar.

A13.5   FINAL PLUG BLASTING, FOSSVATN 01-03 APRIL, 1987 (FIG. A13.5)

It was planned to empty the hollow below the plug for water during the loading operation and the removal of the scaffolding.

Problems with poor electric contact to earth and risk of spark-over made it necessary however to remove all electric equipment and wiring. No live cables were on the inside of the gate.

The result was a hollow full of water. A plastic dinghy was used to reach the face. This caused certain delay during the loading operation.

The loading operation from the top of the scaffolding went well despite of heavy water leakage.

The dismantling of the scaffolding was also delayed on account of the water in the hollow.

The gate was in position on Friday 3 April at 0300 hours and the water filling operation could start.

Work progressed according to plan. The water level started to rise in the shaft and the pressure increased in the air pocket below the plug.

Minor air leakage became evident, however, as the water level increased.

When the water level in the shaft reached level 568, the air leakage was greater than the capacity of the compressors. In addition, considerable water leakage was observed through the gate. The leakages were in the joints between the separate parts of the gate and caused by a fault in the design of the gasket system.

These leakages had by now created a critical situation. 32 m of waterfilling in the shaft still to go, corresponding to an increase in the air pressure at the face of 3.2 bar. This would in turn lead to even higher leakages and the risk of loosing the entrapped air all together.

Blasting of the plug directly into the water would create extremely high pressure on the gate construction and most likely lead to damages.

New calculations were undertaken. Water filling to level 572.8 would give an optimal relation between the water level in the shaft and the entrapped air.

The calculations showed that the upsurge in the shaft would reach 10 cm above the floor level in the gate house and that the strain on the gate would rise close to the maximum allowed. This was accepted by the construction project management.

The uneven and irregular tunnel wall would to a certain extent damp the pressure wave and the upsurge in the shaft. This gave a small safety margin in relation to the calculated values.

The last meters of water was filled up. The compressors were running at full speed to compensate for air leakage.

Level 572.8 was reached and water and air supply shut off, while preparations were made for the shooting.

The final plug was blasted at 20.08 hours on 3 April.

After another 2 minutes the flag, which was placed above the break-through and some 400 m off shore, started to move upwards. Then the 1.2 m thick layer of ice on the Fossvatn lake burst.

Another successful plug shooting of the Kobbelv project was completed.

The gate withstood the pressure and the floor of the gate house was dry after the blasting. The measurements showed somewhat less pressure against the gate and 20 cm less upsurge in the shaft as compared with the calculated values.

A13.6 THE BREAK-THROUGH

	Calculations	Results
Size of blast	270 cu. m	
Number of bore holes, 45 mm	159	
Number of boreholes, 100 mm	8	
Quantity of explosives	1200 kg	1100 kg
Specific consumption of explosives	4,50 kg/m^3	4,00 kg/ m^3
Number of detonators	318, two in each hole	
Quantity of filled up water	Approx. 20,000 m^3	Approx. 19,000 m^3
Quantity of filled up air	6,150 Nm^3	3,200 Nm^3
	620 m^3 (9,9 bar)	420 m^3 (7,5 bar)
Max. air pressure after water filling	9,9 bar	7,5 bar
Max. air pressure on firing	14,3 bar	12,5 bar
Relevant levels:		
– Water level Fossvatn lake	620,0	620,0
– Floor, gatehouse	624,3	
– Sill at gate	501,0	
Level, water filling in shaft	600,0	572,8 *
Upsurge in shaft		624,2 *
Upsurge in shaft		51,4 m *
Pressure against gate:		
Max. allowed	18,1 bar	
Max. pressure against gate on firing	17,7 bar	17,3 bar

* Reduced water filling

APPENDIX 14

Oseberg transportation system – break-through to the sea

A14.1 INTRODUCTION

Crude oil from from the Oseberg field in the North Sea is transported through pipelines for storage in rock caverns at Sture, 65 km north of Bergen.

The pipeline is placed in a tunnel close to the shore line. The tunnel is 2.3 km long and the tunnel opening is at a depth of 80 m.

The pipeline was after the plug blasting hauled directly from the support vessel and into a lock established between two concrete plugs and hooked up to the pipeline in the tunnel (Figs A14.1 and A14.2).

Norsk Hydro A.S. was responsible for the transportation system project with Astrup Høyer A.S. as contractor for the tunnelling.

A14.2 PRELIMINARY INVESTIGATIONS

Different methods of transporting the crude through the shore line zone were considered. The alternative which finally and clearly appeared to be the best was a solution with a tunnel with break-through into the sea from the tunnel side.

An extensive survey of the shoreline zone was made to find the best place for the tunnel and location of the break-through.

Seismic surveys were made and core drillings undertaken from islets and skerries. ROV (Remotely Operated Vehicle) was to a great extent used to videofilm the seabed at alternative locations. Loose rock mass and sediments were recorded by acoustic measurements.

A14.3 GEOLOGY

The rock in Øygarden is precambrium, consisting of granite-gneiss with bands of amphibole, pegmatite and quartz layers. The rock is not very fractured outside the recorded weak zones. The bedding of the rock is slightly declining in an northern direction. This was favourable with respect to leakage. The leakage was very

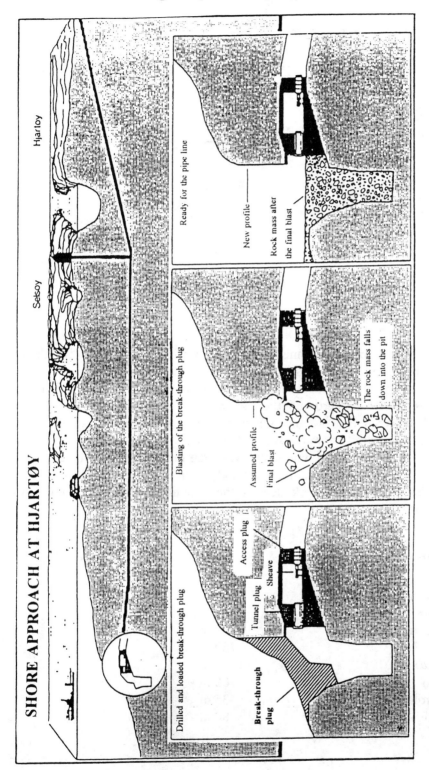

Figure A14.1. Shore approach at Hjartøy.

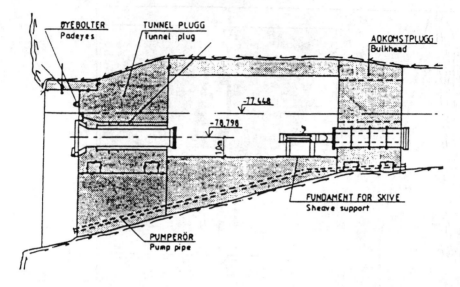

Figure A14.2. Concrete stop plugs and lock.

closely controlled during the tunnelling operation primarily by extensive and detailed inspections and good working procedures for the tunnel work.

The seismic refraction measurements and the topographic conditions indicated several zones of weakness crossing the tunnel.

The strait between Hjartøy and Selsøy indicated a major zone of weakness. Core drilling in the strait showed a zone 40 m wide.

The tunnel was to be filled with water when the pipeline was in place. This, in addition to the presence of swelling clay and the need for good working conditions to achieve high quality work, decided the amount of sealing and stabilization work required.

A14.4 SUBSEA TUNNELLING

The chosen track for the tunnel:

Cross-section	26,5 m^2
Total tunnel length	2375 m
– Decline 1:7	760 m
– Climb 1:20	1440 m
– Climb 1:6	100 m
– Break-through approach	35 m
Water depth at break-through	80 m

Average overburden	60-70 m
– Hjartøysundet	50 m
– The outer section	16 m
– At break-through	5-6 m

Contractors equipment for the tunneling:

Drilling rig:	Strømnes, with 3 drilling units and loading cage
Loading and transport:	Cat 980 C and Volvo 4600, trucks
Concrete casting:	Pump truck and scaffolding
Spray-on concrete:	Robocon A.S. – complete rig on truck

Certain drilling and loading problems were encountered in the declining section (1:7) caused by clay in vertical cracks perpendicular to the tunnel direction.

The greatest challenge to the tunnelling was the crossing of the Hjartøy zone. Here two sections of the tunnel were strengthened by cast concrete linings at the face for a length of approx. 70 m.

The main part of the tunnel was blasted with 2 m rounds at a time. Fibre reinforced spray-on concrete or shotcrete was applied immediately after loading followed by full concrete casting. The area was later strengthened by a reinforced concrete construction.

Standby arrangements and procedures to handle major water leakages were in place from the very start.

Pumps with capacity of 1500 l/min were installed. Later the pumping capacity was increased to 3000 l/min and to 6000 l/min prior to work in the break-through area.

Drilling through into the sea with a borehole diameter of 54 mm resulted in a major leakage of approximately 4000 l/min.

When the blasting short of the plug was completed, the leakage into the tunnel was approx. 200 l/min. At the tunnel face at the plug the seepage was somewhat below 20 l/min.

The work procedure for the entire tunnel included continous test drilling to prepare for unexpected and large water leakage. 3-32 m deep test holes were drilled for every 24 m of tunnel within 130 m of the plug. For the remaining part of the tunnel the test drilling was intensified with 6-24 m deep holes and 2-6 m deep holes for every 16 m of tunnel.

Maximum advance for the entire tunnel length was 92 m/week Average advance for the first 2200 m of tunnel was 40 m/week.

Stabilization work performed

Full concrete lining	275 m
Number of rock bolts	5000
Fiber-reinforced spray-on concrete	1800 m^3
Injection, (hours at the tunnel face)	200 hr.
Injected material – concrete based	20 tons

The rock condition in the last 1000 m of tunnel was good and better than expected. Nevertheless, the last 130 m of the tunnel was driven with reduced round length and for the last 20 m also with a cross section divided in two parts (Fig. A14.3).

Great emphasis was placed on cautious blasting to avoid unnecessary cracking of the rock in the area of tunnel break-through and in the area where the two concrete plugs were to be poured (Figs A14.1 and A14.2).

The vibrations generated by every blast in the area were monitored in order to record ease of breakage.

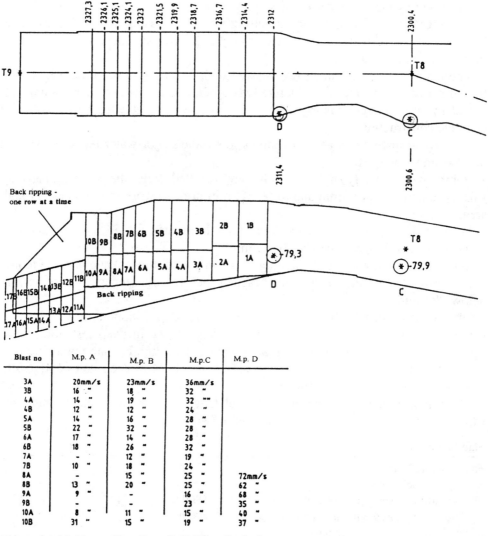

Blast no	M.p. A	M.p. B	M.p.C	M.p. D
3A	20mm/s	23mm/s	36mm/s	
3B	16 "	18 "	32 "	
4A	14 "	19 "	32 ""	
4B	12 "	12 "	24 "	
5A	14 "	16 "	28 "	
5B	22 "	32 "	28 "	
6A	17 "	14 "	28 "	
6B	18 "	26 "	32 "	
7A	-	12 "	19 "	
7B	10 "	18 "	24 "	
8A	-	15 "	25 "	72mm/s
8B	13 "	20 "	25 "	62 "
9A	9 "	-	16 "	68 "
9B	-	-	23 "	35 "
10A	8 "	11 "	15 "	40 "
10B	31 "	15 "	19 "	37 "

Figure A14.3. Tunnelling from P 2312 to final plug.

After blasting all the way to the plug, the water seepage in the break-through area was on the lower side of 20 l/m. This confirms that the blasting had been performed cautiously and with great precision.

Figures A14.3-A14.8 show the blasting plans with reduced round lengths and with a cross section split two ways.

A14.5   SEALING OF TEST HOLES AFTER DRILLING THROUGH INTO THE SEA

Drilling through into the sea started 30 m short of the break-through area. The holes were sealed as far out towards the sea as possible by specially designed ex-

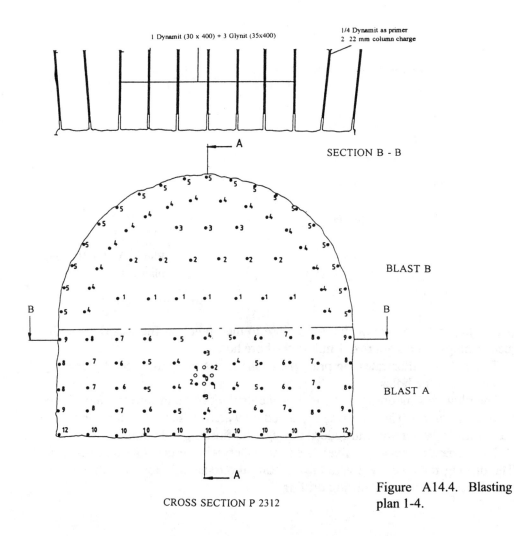

CROSS SECTION P 2312

Figure A14.4. Blasting plan 1-4.

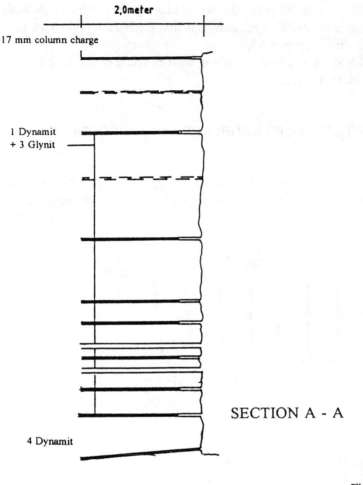

2,0meter

17 mm column charge

1 Dynamit
+ 3 Glynit

SECTION A - A

4 Dynamit

SECTION P 2312

Figure A14.5. Blasting plan 1-4.

pandable plugs. Sealing of the holes close to the tunnel profile was done with injection plugs placed approx. 1 m into the bore hole.

Figure A14.9 illustrates the principle of the hydraulic plug specially developed for sealing of the holes.

The plug was mounted at the end of the drilling rod and inserted into the hole by the drilling rig. The plug was expanded by waterpressure from a waterpump on the rig itself. Water was introduced through the flushing hole of the drilling rod.

Two separate one-way valves keep the waterpressure constant and permanent. The drilling rods were extracted by rotating the rods anti-clockwise, i.e. the normal direction of rotation during drilling.

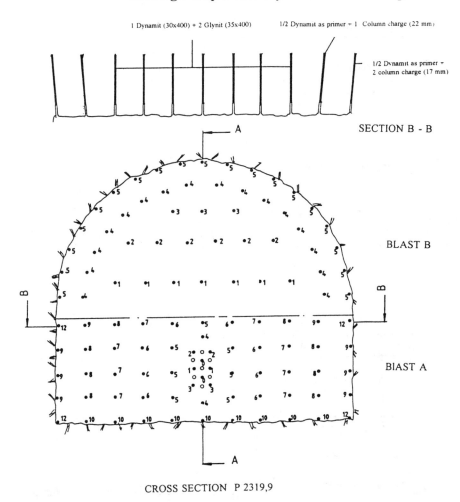

1 Dynamit (30x400) + 2 Glynit (35x400)   1/2 Dynamit as primer + 1 Column charge (22 mm)

1/2 Dynamit as primer +
2 column charge (17 mm)

SECTION B - B

BLAST B

BLAST A

CROSS SECTION P 2319,9

Figure A14.6. Blasting plan 5-10

Thereafter, the hole was injected with a concrete based mortar and finally (2 m) with expandable mortar.

The drilling through into the sea was done systematically and stepwise prior to and during the blast drilling.

Totally, 60 holes were drilled through into the sea for detailed survey of the rock surface.

A14.6 THE BREAK-THROUGH BLAST

The main tunnel was first driven all the way into the break-through area followed by trimming for the break-through blast by light back ripping. Thereafter, a pit

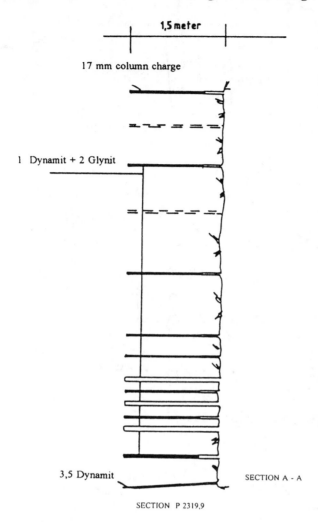

1,5 meter

17 mm column charge

1 Dynamit + 2 Glynit

3,5 Dynamit

SECTION P 2319,9

SECTION A - A

Figure A14.7. Blasting plan 5-10.

tunnel was driven with the same degree of caution as observed in the last part of the main tunnel. A shaft from the pit tunnel to the break-through area was blasted after the drilling of the plug. Figure A14.10 shows the final shape of the shaft and break-through plug.

The break-through blast had a shape not unlike the cutting of a tunnel. The top of the cutting was stitched together with bolts prior to the drilling of the plug. The face was reinforced with glassfibre bolts and sprayed-on concrete.

The break-through plug was drilled with horizontal and vertical holes so that a wedge cut was established with a breakout outwards and away from the concrete plugs. The longest horizontal holes were approximately 20 m deep, the vertical holes approx. 15 m.

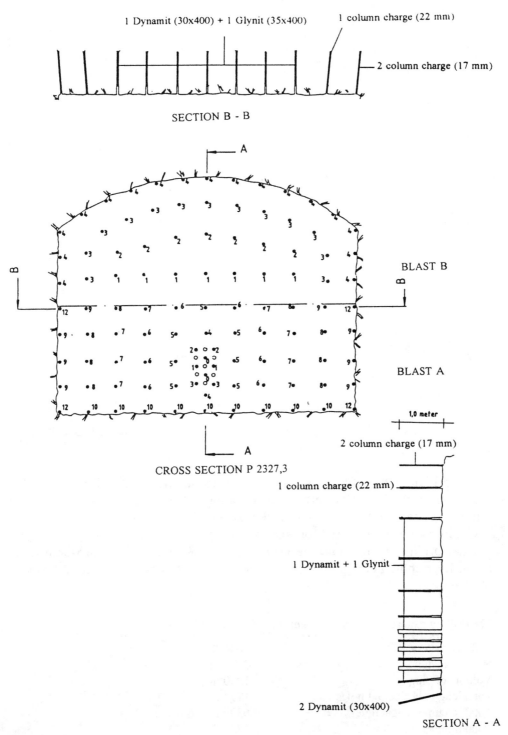

1 Dynamit (30x400) + 1 Glynit (35x400)

1 column charge (22 mm)

2 column charge (17 mm)

SECTION B - B

BLAST B

BLAST A

1,0 meter

CROSS SECTION P 2327,3

2 column charge (17 mm)

1 column charge (22 mm)

1 Dynamit + 1 Glynit

2 Dynamit (30x400)

SECTION A - A

Figure A14.8. Blasting plan 11-17.

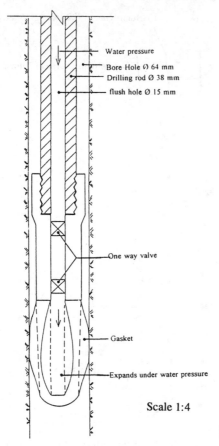

Water pressure

Bore Hole Ø 64 mm

Drilling rod Ø 38 mm

flush hole Ø 15 mm

One way valve

Gasket

Expands under water pressure

Scale 1:4

Figure A14.9. Principle of the hydraulic plug.

The drilling pattern for the vertical and horizontal holes was 70 × 70 cm. The holes were drilled 50 cm short of break-through. The plug was drilled with the drilling rig used for the tunnelling work and with 2½" drill bits.

Guiding sleeves were used for boreholes deeper than 6 m.

After drilling of the plug, the water leakage in the area was measured to 18 l/min. Every hole in the plug was lined with plastic tubing with internal diameter of 50 mm.

The drilling pattern is shown on Figures A14.12 and A14.13

Drilling pattern	70 × 70 cm
Hole diameter	2½ inch
Volume of plug	1240 m^3
Total length, horizontal holes	1797 m
Total length, vertical holes	630 m
Total number of holes	222

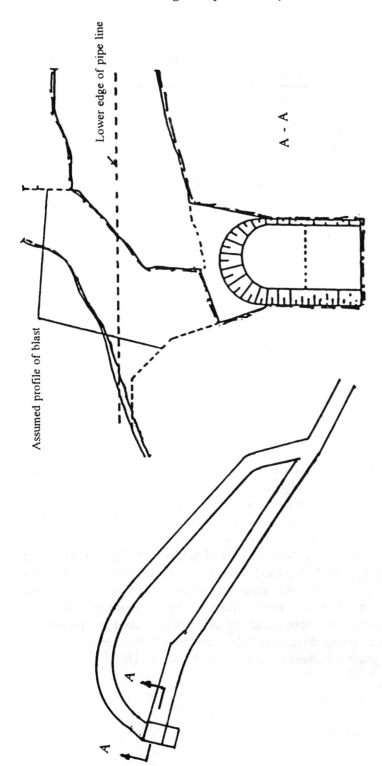

Figure A14.10. Shaft and pit tunnel.

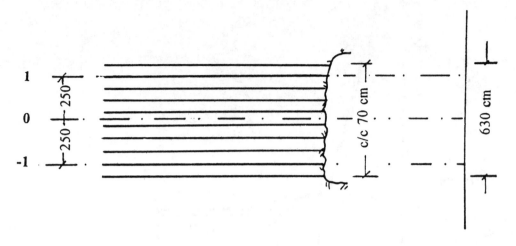

PLAN VIEW (SECTION, ROW 1 - Q) A - A

DRILL BIT: 2,5"

DRILLING PATTERN (HORIZONTAL AND VERTICAL): 70 CM X 70 CM

EVERY HOLE TO BE DRILLED TO 50 CM SHORT OF BREAK THROUGH

Figure A14.11. Final plug – drilling.

The break-through plug was drilled prior to the blasting of the shaft pit and before the pouring of the concrete plugs.

Millisecond delay detonators, Group 1, were used with reinforcing sleeves over the leads and with intervals from 1-16 and 20 m long leading wires. Two detonators of the same number went into each and every hole.

The vertical holes were loaded with cartridged explosives, (35 × 600). Extra Dynamit and detonators placed in plastic tubes of inside diameter 40 mm. The plastic tubes with the explosives and detonators were pushed upwards into the boreholes and locked in place by wooden dowels. The horizontal holes were loaded in a regular manner. Detonators and explosives were tested and guaranteed to withstand a waterpressure of 90 m over 7 days by Dyno Industrier A.S.

Loading and firing plans are given on Figures A14.14–A14.16.

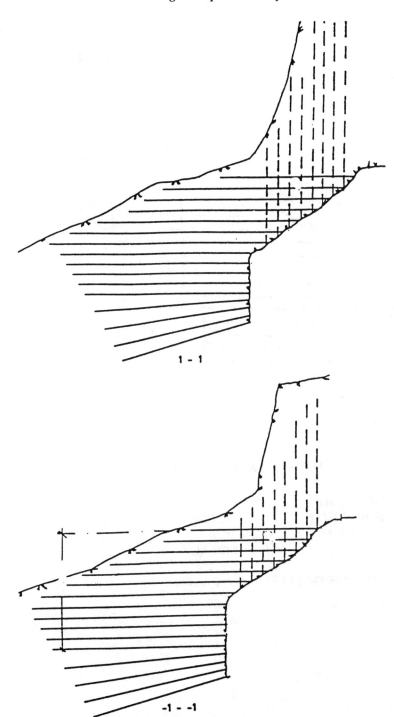

1 - 1

-1 - -1

LONGITUDINAL SECTION

Figure A14.12. Final plug – drilling.

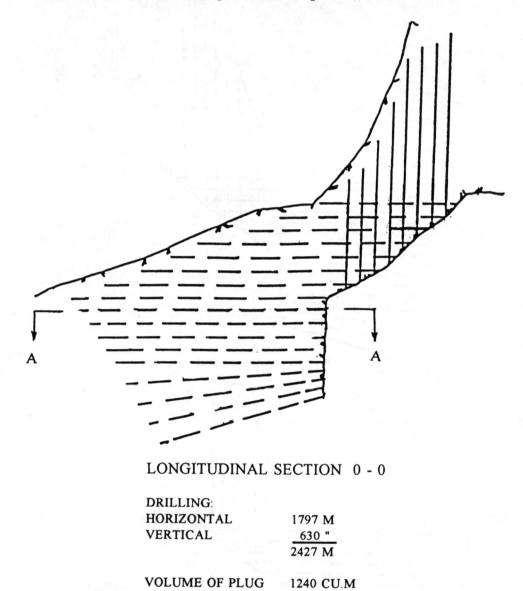

LONGITUDINAL SECTION 0 - 0

DRILLING:
HORIZONTAL   1797 M
VERTICAL   630 "
                    2427 M

VOLUME OF PLUG   1240 CU.M

Figure A14.13. Final plug – drilling.

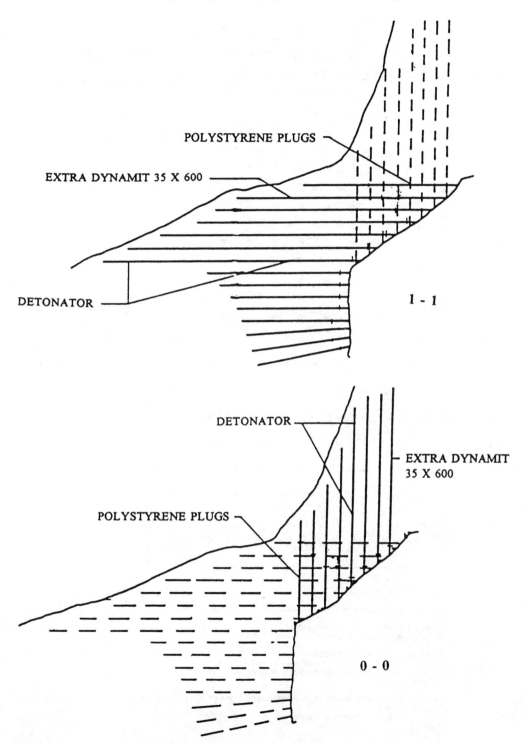

Figure A14.14. Final plug – loading and ignition.

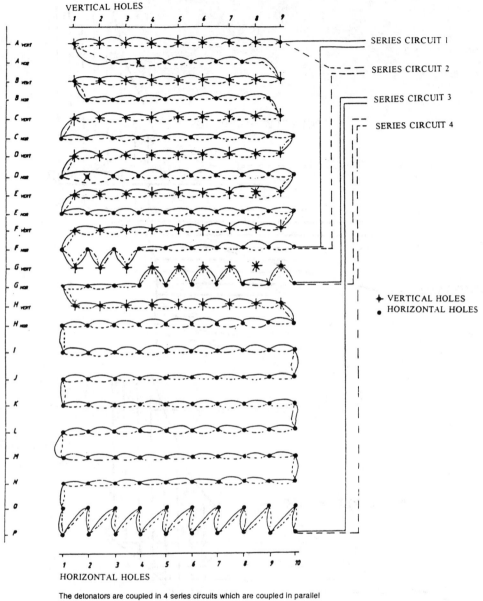

VERTICAL HOLES

SERIES CIRCUIT 1

SERIES CIRCUIT 2

SERIES CIRCUIT 3

SERIES CIRCUIT 4

✦ VERTICAL HOLES
● HORIZONTAL HOLES

HORIZONTAL HOLES

The detonators are coupled in 4 series circuits which are coupled in parallel circuit.
The top detonators are coupled to series circuits 1 and 2
The bottom detonators are coupled to series circuits 3 and 4

The resistance of the 4 series circuits must be equal, max diviation +/- 1 %
The leading wires not to be cut , and used with the length supplied.
Surplus leading wires between the boreholes to be wound up in bundles
to get a better view.

The coupling of the leading wires to be done with Scotchlock type UR
Use plyer type E9.Y
The hook-up to the firing cable to be insulated with tape and sealing
compound.

Figure A14.15. Final plug – hook-up pattern.

4 - FIRING CABLES 2 X 2,5 SQ.MM FOR SERIES CIRCUITS 1 - 4
TO BE PULLED THROUGH THE TWO CONCRETE PLUGS

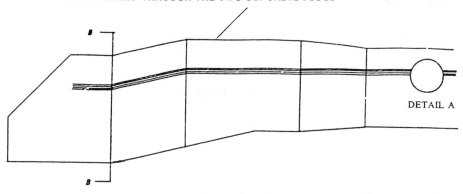

DETAIL A

2 - FIRING CABLES 2 X 2,5 SQ.MM TO BE HOOKED UP
IN PARALLEL CIRCUIT AND STRETCHED TO THE POINT OF FIRING

CRIMPING SLEEVE FOR WATERTIGHT INSULATION

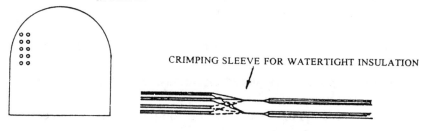

PARALLEL CIRCUIT HOOK-UP

B - B

DETAIL A

NUMBER OF DETONATORS	444
NUMBER OF DETONATORS PER SERIES CIRCUIT	111
RESISTANCE PER SERIES CIRCUIT 111 X 3,6 OHM	400
RESISTANCE SERIES / PARALLEL CIRCUIT HOOK-UP	100
RESISTANCE FIRING CABLE (1,4/100X2500/2)	17,5
TOTAL RESISTANCE	117,5 OHM
EXPLODER: DYNOCON 30 OR EQUIVALENT	
VOLTAGE	1100 VOLT
CURRENT 1100/117,5	9,4 AMPERE
CURRENT, PARALLEL CIRCUITS 9,4/4	2,35 AMPERE

DETONATORS: GERMAN MADE MILLI-SECOND GROUP 1 WITH
                20 M LEADING WIRES, REINFORCED WITH SLEEVES

Figure A14.16. Final plug – hook-up pattern.

A14.7 SHOCK PRESSURE AGAINST THE CONCRETE CONSTRUCTION VIBRATIONS

A total of 2400 kg of explosives divided between 16 intervals was fired in the blast. The greatest charge per interval was on detonator No. 16 with 225 kg distributed between 15 holes.

Limiting value for the particle velocity was set at 450 mm/s and with a max. amplitude of 1500 µm.

The concrete plug and steel gates were designed for a shock pressure of 25 bar.

'Fly rock' from the blast was anticipated against the concrete plug and the steel gates and good coverage was required to protect the construction.

Based on calculations and evaluations, it was decided to pre-compress the air-pocket in the shaft and the pit to 4.5 bar. Peak pressure was calculated to 25 bar without any damping effect and to 17.5 bar with damping due to friction during inflow. A 16 m^3 compressor, placed on the outside of the entrance to the tunnel, and 3" tubing in the tunnel were used for the compression work. The time for compression was estimated to be 24 hours. The volume of the shaft and the pit below the plug was 3500 m^3.

A vibration recorder was placed on the inside of the outer plug and measured the entire vibration sequence.

Two sets of hydrophones were installed on the outside of the concrete plugs in the main tunnel and in the pit tunnel.

The blast was fired 16 March 1987.

The following recordings were made:

	Main tunnel	Pit tunnel
Vibrations		
Particle velocity	280 mm/s	
Amplitude	700 µm	
Gas pressure	12.5 / 11.5 bar	15.0 / 14.3 bar

The recorded peak pressure is below the calculated values. The recorded and calculated values are in good agreement when the damping factor is taken into account.

Damage or leakage of any consequence were not observed. The total leakage in the area increased from 20 to 70 l/min.

Inspection with ROV and video camera revealed a full opening. The protection over the gates had disappeared, but the concrete plug and the gates itself were undamaged.

The side walls and the back wall were after blasting as intended with good contours and without need of scaling and stabilization.

A few plastic tubes were found with undetonated explosives. It is anticipated that this was caused by a few of the deep horizontal holes being cut off prior to detonation of the hole.

The scaling operation and preparatory work for the hauling of the pipeline into the entrance of the tunnel took three days with support vessel.

A14.8 QUALITY CONTROL

The consequences of misjudgement and insufficient testing, planning and risk evaluation can prove very costly and dangerous and cause interruption of progress which in turn may lead to delays, rock slides and in-flow of water.

Safeguards against poor judgement, lacking plans and procedures are critical elements in this type of construction work.

Every operation was carefully planed in detail and adjusted well ahead of time when unforeseen problems were encountered.

Important elements in this connection were:

– Systematic test- and core drilling, and prompt sealing of leakages according to worked out procedures in areas where problems could be expected.

– Cautious blasting with shorter rounds and reduced charges.

Leakage in the break-through area where the water pressure was 80 m and where the over burden was reduced to 5 m, was only 20 l/min. This indicates that the blasting was done cautiously and in a controlled manner.

The author

John Johansen, the author, is a B.Sc. graduate from the Bergen Higher School of Technology in 1950 in the field of civil engineering.

1950-1961. With the Norwegian Harbour Authority, project planning and construction works management.

1961-1990. Dyno Industrier, Explosives Group, initially as a technical service engineer, later as manager of Dyno Consult.

1990-present. Independent consultancy service in the field of rock blasting and blast design.

John Johansen has been actively involved as a consultant in a number of difficult blasting operations in Scandinavia and abroad including Alaska, Holland, Chile and most recently in India.

John Johansen is the author of numerous publications covering surface-, underground- and offshore blasting operations, including plug shooting and demolition work.

Supplements

Improvement of tunnel profile by means of electronic detonators

ROLF KÖNIG
Dynamit Nobel, Troisdorf, Germany

1 FOREWORD

Some of the most significant landmarks in technological progress are represented by the achievements in the field of energy control. This is particularly true concerning explosives. Let us consider in fact the increasingly sophisticated applications and the more and more important results obtained and we shall understand that such a progress is mainly due to the development in priming devices (detonators), whereby it is possible to control with greater and greater precision that the explosive energy content be transformed in work.

Over the past few years Dynamit Nobel developed a detonator of the electronic type (Dynatronic) offering a performance at a level quite superior with respect to either the electric detonator or the non-electric shocktube detonator presently available on the market.

Such a high performance is determined by:
- Extreme precision in delay time which ensures the simultaneous explosion of all detonators with the same delay number.
- The possibility to decide and select just before firing the time interval between a given and the preceding delay number.

Therefore, when using electronic detonators it is possible to plan the delay interval which, according to the type of round, permits to obtain an optimal fragmentation of the blasted material and to substantially reduce the vibration induced by blasting.

Moreover, effective simultaneity in the firing of groups of shots allow to improve the excavation profile and this is particularly appreciated in tunneling, where the narrow limits imposed for the overbreak together with the relevant costly penalties make it compulsory to change radically the blasting techniques normally adopted. This paper illustrates the excellent results in the driving of a tunnel where the contour was considerably improved by adopting Dynatronic detonators to fire the perimeter shots with exactly the same firing time, i.e. simultaneously, without any scattering.

2 THE ELECTRONIC DETONATOR (DYNATRONIC)

The time elapsing between the priming of electric detonators (or non-electric ones) and their explosion is a function of the length and burning speed of the delay element inserted next to the detonator's primary charge (cf. Figs 1 & 2). However careful the quality of the production of such elements may be, absolute simultaneity within the same round in the explosion of all detonators identified by the same delay number is never guaranted. The detonators which are normally used in tunnels feature delay intervals of 100 ms or 250 ms and the difference between actual delays can be as much as 10-20 ms or even more.

On the contrary, the electronic detonator avails itself of a series of coded impulses transmitted to capacitor through a microchip which constitutes the delay element. The capacitor is such as to produce a discharge which ignites the fusehead. The electronic components which determine the designed delay time permit the manufacture of detonators where the difference between the nominal delay time and the actual delay is smaller than 0.5 ms.

Moreover, the time interval between delay numbers is not factory made but determined at the time of use, with the possibility of making the selection of constant delay intervals in the range 1-100 ms or variable intervals according to the type of blasting machine used.

Dynatronic detonators, which are available in 61 delay numbers (0-60), offer numerous different delay times, as shown at Table 1.

Also as far as safety is concerned, it is out of question that the electronic Dynatronic detonator is markedly superior to the electric detonator. The ignition of this electronic detonator can take place only through a series of coded signals which cannot be repeated accidentally, nor are they naturally present on earth. In order to be activated an electronic detonator needs an electric current of adequate intensity together with a very special electronic access key.

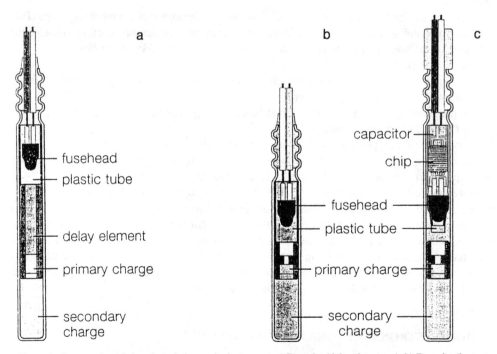

Figure 1. Construction of electric and electronic detonators: a) Dynadet (delay detonator), b) Dynadet (instantaneous detonator) and c) Dynatronic (delay detonator).

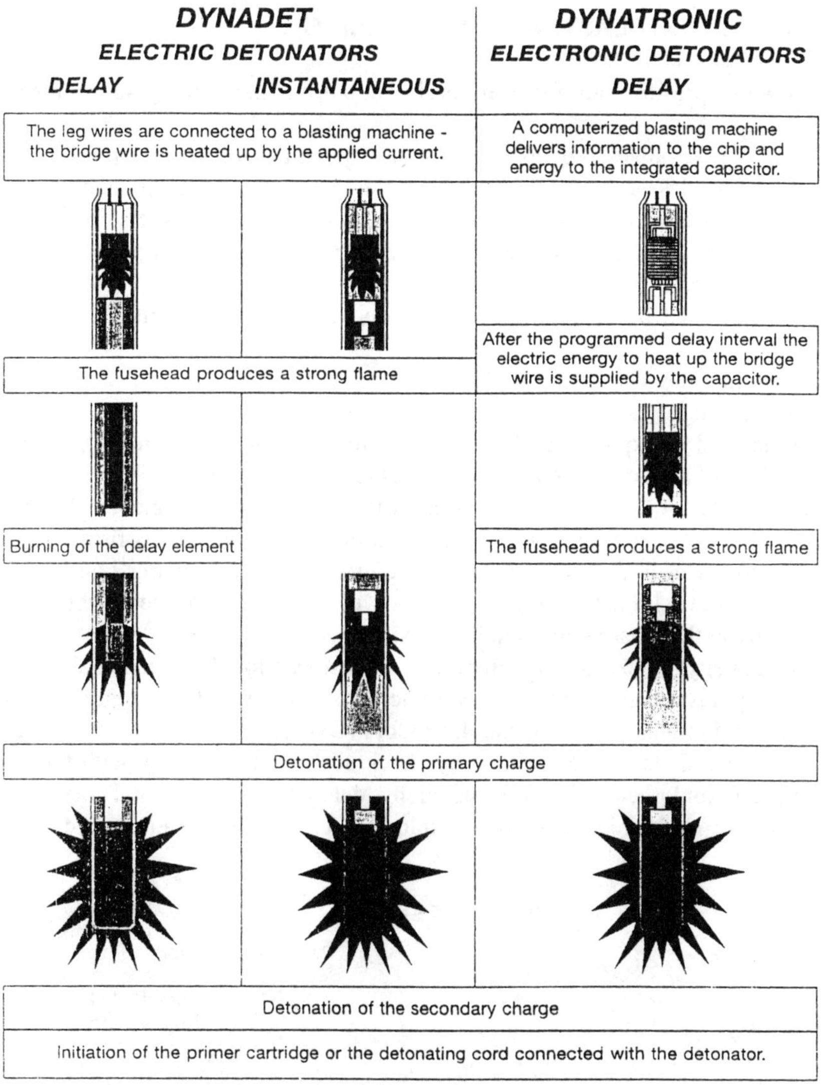

DYNADET ELECTRIC DETONATORS		DYNATRONIC ELECTRONIC DETONATORS
DELAY	**INSTANTANEOUS**	**DELAY**
The leg wires are connected to a blasting machine - the bridge wire is heated up by the applied current.		A computerized blasting machine delivers information to the chip and energy to the integrated capacitor.

The fusehead produces a strong flame		After the programmed delay interval the electric energy to heat up the bridge wire is supplied by the capacitor.
Burning of the delay element		The fusehead produces a strong flame

Detonation of the primary charge

Detonation of the secondary charge

Initiation of the primer cartridge or the detonating cord connected with the detonator.

Figure 2. Operation of electric and electronic detonators.

Table 1. Firing times of Dynatronic detonators.

Delay on blasting machine (ms)	Delay number (detonator)					
	0	1	2	58	59	60
	Delay time (ms)					
1	0	1	2	58	59	60
2	0	2	4	116	118	120
3	0	3	6	174	177	180
55	0	55	110	3190	3245	3300
98	0	98	196	5684	5782	5880
99	0	99	198	5742	5841	5940
100	0	100	200	5800	5900	6000

3   PROFILE (PERIMETER) BLASTING IN TUNNELLING

In all excavation works profile is an important problem, but it is particularly so in tunnelling today.

A good profile implies smaller overbreak, smaller quantities of blasted rock to be loaded and carried away, lesser need of concrete for the lining. Moreover, a good profile implies that the fractures induced into the rock by blasting of the contour shots do not exceed design limits, thus restricting the operations of supporting, spraying, and waterproofing to the bare necessary.

The contour is improved by designing the blast, and especially the contour shots as follows:
– Making sure that holes are carefully parallel
– Narrowing hole spacing
– Using detonating cord or explosives of a type specially suitable for the result wanted, loaded in cartridges of a diameter substantially smaller than the hole.

Blasting theory provides another useful hint, perhaps the most important one, to obtain a profile coinciding as much as possible with design profile, i.e.: simultaneous firing of all the contour shots. However, in tunnel rounds the contour shots are to be primed with high delay numbers which fire with substantial scattering. Only the electronic detonators provide real simultaneity in the blasting of contour shots.

Since this simultaneity of explosion is required exclusively for the contour shots, it may be economically advantageous to use electronic detonators, which are more expensive than electric and non-electric shocktube detonators (Dynashoc), exclusively for contour shots. All the other shots may be primed by shocktube detonators, each with the required delay number, and their tubes connected by detonating cord to a Dynatronic detonator zero delay number which starts their explosion. The zero delay electronic deto-

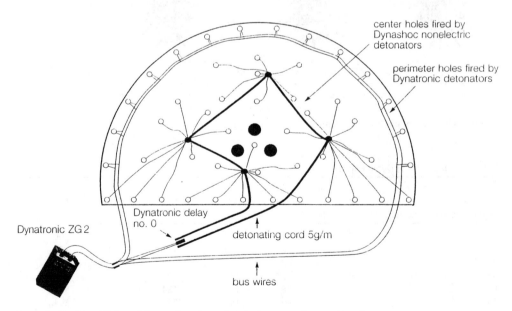

Figure 3. Combined firing of electronic and nonelectric detonators in tunnel heading.

nator is connected, in its turn, through a bus line, to the electronic detonators in the contour shots, which will explode simultaneously, but with convenient delay with respect to the last Dynashoc delay number (cf. Fig. 3).

4 APPLICATION OF ELECTRONIC DETONATORS

As an interesting example of excellent profile obtained by means of Dynatronic detonators, the Mitholz access tunnel can be mentioned. This tunnel, which is a part of the recent plan for railroad crossing through the Swiss Alps, is 1550 m long. As the excavation proceeded the front met slate and limestone alternately.

The geomechanical features of the rock were quite poor, in fact the frequent slides in hole walls made it difficult to load the explosive cartridges.

The general plan of the work had envisaged two concrete linings of the tunnel, one to be applied immediately after excavation, the second, and final one, to be applied much later and perhaps by a different contractor.

As a precaution against the risk that at the time of the second lining the need for concrete might be greater than planned, at the beginning of excavation the organization awarding the job set forth a penalty referred to the amount of overbreak and tied to the cost of the concrete used. Such penalty was 300 Swiss Francs (CHF) per cu. m. Since the tunnel perimeter amounted to 21,3 m, the cost per every meter of advance was 63,90 CHF per centimeter of overbreak.

At the beginning of the excavation (carried out by rounds primed by HU electric detonators) the average overbreak was 25 cm and the relevant penalty was 63.90 × 25 = 1597,5 CHF per meter of advancement, the contractor decided to try Dynatronic electronic detonators according to the procedure illustrated in the previous section.

The profile was greatly improved, and it improved further after the workers became familiar with the new technology.

A check of the profile was carried out after each round and the results are given in Table 2.

Obviously, such results could not be obtained without the help of an adequate drilling machine. Atlas Copco Boomer 353 with Bever Data Control for computerized drilling was the one selected, its most interesting features being:
– Extreme precision,
– Possibility to drill, by the help of a computer, a designed round pattern,
– Check the drilling during operations with possibility to correct possible mistakes.
The characteristics of the round are given below and in Table 3 and refer to the drilling pattern of Figures 4 and 5.

Table 2. Overbreak reduction.

Round position (m)	Average overbreak (cm)
100/200	25
200/300	14
300/400	13
400/500	12
500/600	11
600/700	10

Table 3. Characteristics of the round.

Section	65.75 m²
Cut	V-Cut
Hole diameter	51 mm
Hole number	137
Average hole length	4.00 m
Total length drilled	548 m
Average advance	3.80 m
Volume blasted	250 m³
Explosive used	450kg
Detonating cord (80 g PETN/m)	110 m
Dynashoc LP (0-25)	108 pcs
Dynatronic (0-28-30)	30 pcs

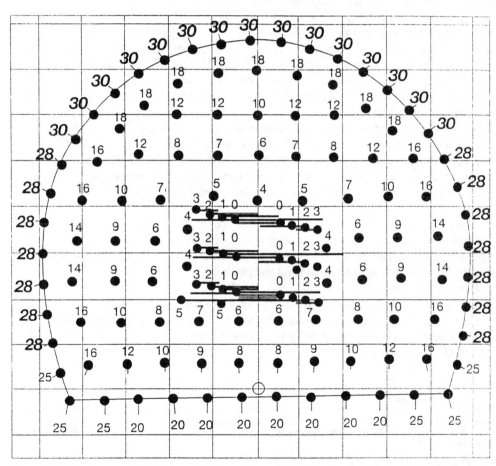

Figure 4. Firing pattern of Mitholz tunnel. Firing a tunnel round with electronic and nonelectric detonators. Nonelectric Dynashoc 100 ms delay detonators. No. 0 to no. 25 are initiated by Dynatronic delay no. 0. Perimeter holes are initiated by electronic detonators' delay numbers 28 and 30. The electronic detonators are programmed with 100 ms delay. The resulting firing time is 2800 ms, respectively 3000 ms.

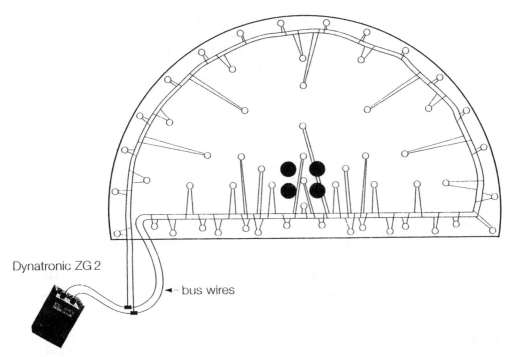

Dynatronic ZG 2

◄─ bus wires

Figure 5. Firing a tunnel round with electronic detonators.

5 CONCLUSIONS

The excellent result obtained (i.e. reduction of the average overbreak from 25 to 10 cm) must now be reviewed from the economic viewpoint, comparing the greater costs which the use of electronic detonators entails to the savings deriving from the reduction of the overbreak and consequently from the reduction of the contract penalty.

Having stated that the number of detonators per round is the same both with electric detonators, and that (at the time when the tunnel was constructed) electronic detonators costed 20 CHF more than electric detonators, the increase in round cost by using Dynatronic was $20 \times 30 = 600$ CHF (since the round uses 30 Dynatronic detonators).

On the other hand, being
- Tunnel perimeter = 21,3 m
- Overbreak reduction = 25 – 10 = 15 cm = 0,15 m
- Round advancement = 3,80 m
- Penalty for overbreak = 300 CHF cu.m.

it is easy to reckon how much is saved by reducing the overbreak by 15 cm: $21.3 \times 0.15 \times 3.80 \times 300 = 3,642.3$ CHF/round.

The result is:
- Saving for reduced overbreak   3,642.3   CHF/round
- Greater cost for electronic detonator   600   CHF/round

                                                        3.042.3   CHF/round

This benefit can also be referred to a meter of linear advance (3,042.30/3.80 = 800.6 CHF/m), or to the cu. m. of useful excavation (800.6 : 65.75 = 12.18 CHF/cu.m.). In the assumption to keep round advancement unchanged (3.80 m) it is possible to figure out the minimum reduction of the overbreak paying for increase in cost due to the electronic detonators. Calling 's' this value: $21.3 \times s \times 3.80 \times 300 = 600$ CHF = greater cost for Dynatronic, s = 0.025 m = 2.5 cm.

Finally, after having evidenced the economic benefit deriving from the use of electronic detonators in the particular context of Mitholz tunnel, it is interesting to point out, as a general rule, an additional benefit: Potential reduction of the stress which the blasting of the contour shots induces in the surrounding rock.

As it is well known, the rock between two contour shots can be cut even through their explosion is not simultaneous. This means that the cut is done by the shot exploding first, and the explosion of the second one, being the rock already cut, is useless. In the latter shot only the hole is exploited, which serves as a guide for the cut. On the contrary, if electronic detonators are used, it is possible to count on perfectly simultaneous priming of the profile shots and, therefore, on the synergy of their explosions and it is possible to decrease the linear loading density, thus reducing the stress level on tunnel walls and, as a consequence, the danger that new fractures may be formed in the rock.

REFERENCES

Stratmann, M. 1996. Moderne Bohr- und Sprengverfahren beim Vortrieb des Mitholztunnels. *Nobel Hefte* 1/2, 31-39.
Antonioli, G. & Berta, G. 1996. *Improvement of tunnel profile by means of electronic detonators*. Gallerie e grandi opere sotterranee. November.

Lundby Tunnel – blasting technique for the next millenium – monitoring of vibration

GÖRAN KARLSSON
Bergsäker, Göteborg, Sweden

The Lundby Tunnel is a project designed to improve the environment for the inhabitants of a community divided by a busy transportation link. By directing the traffic into a tunnel under the community, the environment has been greatly improved for those living and working in the area. The tunnel is approximately 2 km long and stretches 10-20 m under the surface.

Approximately 450,000 m^3 hard rock was blasted using 2200 rounds of explosives. In the process 690,000 kg of explosives and 170,000 ignition devices were used.

Monitoring the vibration generated by the blasting is an important tool in the process of controlling environmental stability, blasting efficiency and technical results. To maximize the positive outcome of these parameters, fast access to good information is crucial.

Today, measuring systems, which can meet future requirements are available. One such system is Ava Tunnel, developed by Penttech Engineering Systems AB and Bergsäker Konsult AB. The Lundby Tunnel was the first project to make use of this system.

The tunnel and connecting road systems was build by the Swedish National Road Administration with Skanska as the contractor.

1 PREPARATIONS

Before commencing the project, a precise risk analysis was carried out based on the Swedish standard SS 460 48 66. All buildings and installations within a certain radius were inspected and given a value for the maximum vibration level based on the type of ground underlying the building, its foundation, material content and type of construction. All information from the risk analysis was fed into a database.

The system for seismographic surveillance used in the project consisted of three main parts; seismographs, a communication system and software for processing and presenting results from each blasting event. Mounted on the buildings surrounding the tunnel fronts are seismographs, type Ava 95 equipped with radio modem. At Bergsäker Konsult, the company responsible for the vibration surveillance, a main computer was installed with Ava NT software, a real time system for continuous measurement. To obtain a general view of the project on the computer screen, an essential part of the system is a map of the area indicating the buildings, streets and the contemplated stretch of tunnel. This map is

131

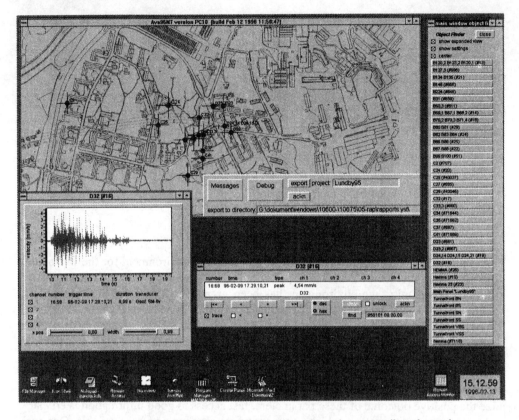

The program map shows a general view of the area enabling the project to be more easily understood.

constantly updated with the positions of the blasting sites and the seismographs recording the blastings.

The Ava NT software processes the incoming recordings, sent from the Ava 95 seismographs by radio, and presents the outcome of the blasting concerning vibrations in the surrounding buildings in reports, which are distributed to the management of the road and tunnel construction.

In order to be able to transmit reports simply and quickly, a report generator was created to include all the information from the risk analysis. The recorded data could then be imported from Ava NT to the report generator. Automatic calculation formulas generating the maximum permitted vibration for each building equipped with a seismograph were created. The calculation included distance from the blasting site and specifics about the building, limiting the permitted vibration level according to the Swedish standard system 460 48 66.

A computer with corresponding software was placed at The Swedish National Road Administration's site office.

2   IMPLEMENTATION

When the contractor blasted a round of explosives, this was registered by the Ava 95 seis-

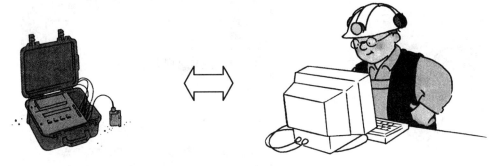

Transmission of vibration data from Ava-95 seismographs to the computer.

mographs giving a complete curve sequence and the peak value for the vibration's vertical speed. The recorded data was transferred immediately thereafter to the main computer.

The indication symbols for the seismographs presented on the main computer's screen changed color from black to yellow to indicate that new data has been transmitted. The new data is transmitted consecutively to The Swedish National Road Administration's computer.

15-30 minutes after the blast, both the peak value and the whole curve sequence can be studied on the screen. The curve sequence shows the vibration generated by each of the ignition devices. This gives you an opportunity to do adjustments in the drilling and charging layout before the next blast.

The whole process was completely automatic, from registering by the Ava 95 seismograph, to several participants being able to see the result on their computer screen at different places. A maximum of approximately 45 measuring points was connected to the system at the same time.

3 REPORTING

The Swedish National Road Administration needed complete measuring data from all the explosions to be included in a report by 10.00 am the next day.

In order to fulfil this, and even be able to report the correct distance between each blasting site and Ava 95 seismograph, Skanska faxed a report every day, latest 9.00 am, containing the number of blast, tunnel, section and time of day. This was then fed into the report generator. Previously, new recorded data had been imported to the report generator. Thus a complete report could easily be delivered before 10.00 am.

Throughout the project a total of 21,810 seismographic recordings were made during a period of 784 days, giving an average of 27 recordings per day.

Storing and processing all data with the aid of a database achieved the following advantages:
– To store information in a structured way
– Unlimited search possibilities
– Fast and varied presentations
– Easy and accurate analysis
The above is valid even 10 years after the conclusion of the project.

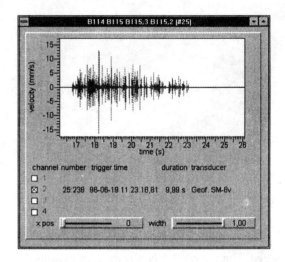

Curve sequence of a tunnel blast.

4 FOUNDATION ANALYSIS

According to Swedish standards, the buildings are given a maximum vibration level based on the type of ground underlying the building, its foundation, material content and type of construction. This is a simplified method to handle the vibration's frequency content, which is crucial for the buildings ability to resist damage.

Around the area of the Lundby Tunnel, the ground underlying the buildings consists mainly of hard rock or clay. For buildings erected in areas where clay and rock meet, discussions arose about what type of ground actually underlies a particular house.

This kind of debate aroused from the fact that the contractor, Skanska, was forced to go under fixed vibration values based on calculations stipulated in the Swedish standard, SS 460 48 66. The maximum vibration value for a building founded on clay is much lower than for a building on hard rock due to the fact that the vibration frequency is much lower in clay than in hard rock, making the vibration more likely to cause damage on buildings.

The implication of a clay-based value of a building in reality founded on rock is a very difficult blasting in order to keep vibration level under the very low clay value.

When Skanska couldn't manage to go under the maximum level stipulated in the risk analysis, they were forced to take action, for example, by shortening the length of each blast. This action, however, costs a great deal of money, hence the discussion about the building being given a wrong (too low) maximum level of vibration.

The solution to the problem was frequency weighted vibration limits. An Ava 95 seismograph was placed on each building in good time before the vibration levels became a problem. The curve sequence was fed to a program for frequency analysis. From the frequency analysis, the limit value then was set.

The result of this action ensured that no expensive blasting action needed to be taken unnecessarily.

5 OBSERVATION

The growing awareness in society of the environmental risks connected to large infrastructure projects will increase the demands for satisfactory control of the surrounding environment in the future. This will have implications for all kinds of building projects.

The key to controlling environmental effects caused by building project is information. Fast access to essential information by as many of the concerned as possible is crucial for the chances of detecting possible problems in time to prevent them from becoming problems.

The new systems, similar to the one used for the Lundby Tunnel project, give us confidence, despite increased demands, in order to carry out explosions in sensitive environments, even in the future.

Intelligent tunnelventilation

SVEIN HAALAND
SIHCon AS, Oslo, Norway

Svein Haaland, the author of this article, is the Managing Director of SIHCon AS, Norway.

For 30 years he has been actively involved in tunnel ventilation projects world-wide.

PROTAN AS is the manufacturer of VENTIFLEX ducting for underground use, and the company use Svein Haaland as their consultant for their export work, which covers more than 30 countries.

1 INTRODUCTION

In the beginning the purpose of underground work was to excavate minerals.

Ventilation was at that time also important to the workers that were exposed to dust, toxic gases and a humid climate below the surface.

Safe and good working conditions is a basic right for people with demanding work where the risk element is higher than for most people in our working society.

Agricola (see Fig. 2) shows his solution to this problem, which was in use 400-500 years ago.

Today, the workers are exposed to many of the same working environmental problems caused by the machines and equipment used inside tunnels and mines. However, better technology enables us to provide a safer and better climate for the workers. Nevertheless, there is room for great improvements in this area, as the available technology is not always used in its most efficient way.

The purpose of a ventilation system is also to give optimum conditions for the equipment in use to ensure that a high production rate is not restricted by substandard working conditions.

2 BASIC CALCULATION OF TUNNEL VENTILATION

a. Airflow required can be calculated based on dilution of toxic fumes:
   $4 \ m^3/min \times kW$ dieselpower is a formula used in many European countries.
b. Minimum velocity in a tunnel should be 0,3 m/s over the full cross-section.

Figure 1. A tunnel project has many variable factors that need to be monitored by a clear mind.

Figure 2. Drawing from Agricolas book on mining about 400-500 years ago.

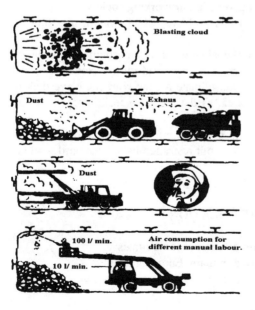

Figure 3.

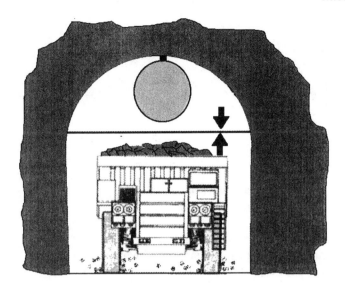

Figure 4.

c. Preheating of the airflow is seldom done even in Norway. The total diesel energy during loading and transport may increase the temperature, even during winter, up to 15°C inside the tunnel.

However, reduced airflow during drilling, scaling and other activities is necessary to avoid frost-problems in the winter.

d. Cooling of the airflow can be necessary in deep mines and tunnels in areas with virgin rock.

The surface temperature of the rock can be up to + 47°C in some cases.

Cooling units, split type, with water cooled condensers have proven to be efficient in such projects.

e. Pressuredrop and selection of duct size and fan type.

Often the calculated pressure drop is based on unrealistic figures.

Limited space for the ductline is a general problem. However, we recommend Nomograms like the one we have enclosed. Here we can see the importance of looking at the energy cost per 100 metres of the actual duct size.

Axial flow fans should be of the high-pressure type with an adjustable pitch or a frequency converter and fixed blades.

A measuring instrument for the airflow and static pressure should always be at the main fan.

3 CALCULATION OF A VENTILATION SYSTEM

When planning a tunnel-ventilation-project we have to remember that the fan and the duct form separate characteristics (see Fig. 5).

Each length of a duct has a separate characteristic as well as each diameter.

These two characteristics must be combined in the same diagram to see the actual airflow from the fan at a different length and with the same diameter of duct. Also, we can compare diameters of duct with the same length.

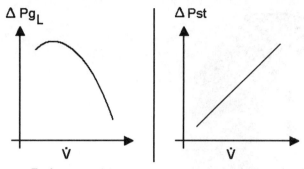

Total pressure of the fan   =   Total pressure consumption of the duct system
Figure 5. Fan and duct characteristics.

AL17 (see Fig. 6) shows the characteristic for a duct diameter ø2m and length of 4 km.

Because the fan has adjustable pitch the capacity can vary form 43-74 m³/s from the fan station.

To calculate the pressure drop in the duct we recommend the VENTIFLEX NOMO-GRAM (enclosed Fig. 7).

Another option is to use a computer-based calculation program. Enclosed you will find the one used in the Qin Ling project (Fig. 8).

4   REVERSIBLE FLOW VENTILATION

Because the airflow in several projects is extremely high in some periods, we have to use new ideas such as reversible flow ventilation.

The blasting fumes are toxic and they should always be removed before the workers re-enter the tunnel.

Since 1987 we have used a mobile fan unit 'Argo-tralle' with a transformer, two fans with dampers and a switch panel which is connected to the main fan on the outside of the tunnel.

This system reverses the airflow and removes the blasting fumes within 20-30 minutes after blasting.

Enclosed drawings show the operating principles (Figs 10-12).

The main advantage of this system is the reduced downtime after a blasting operation.

This system requires a good ductline and a remote control system from the mobile fan.

4.1   *The Aurland- Lærdal tunnel, the world's longest road tunnel – 24,5 km*

The Lærdal side of the project is still under construction by NCC Eeg Henriksen.

This project has an access tunnel of 2000 meters, which leads down to the main tunnel with an inclination of 1:10, continuing with 6000 meters and 7000 meters to respective sides.

Because the tunnelling is driven traditionally with drilling and blasting using large die-seldriven transport units, the ventilation system is a challenge.

The maximum duct diameter that could be used was Ø 2m duct, because of limited clearance.

2 x AL 17 with adjustable blades

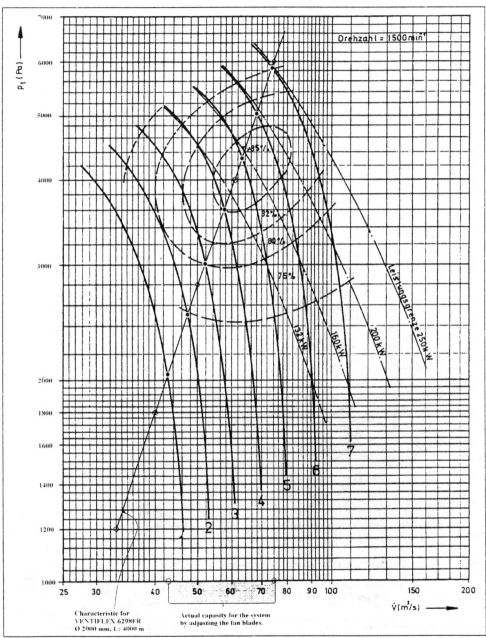

Figure 6. Monogram for calculations.

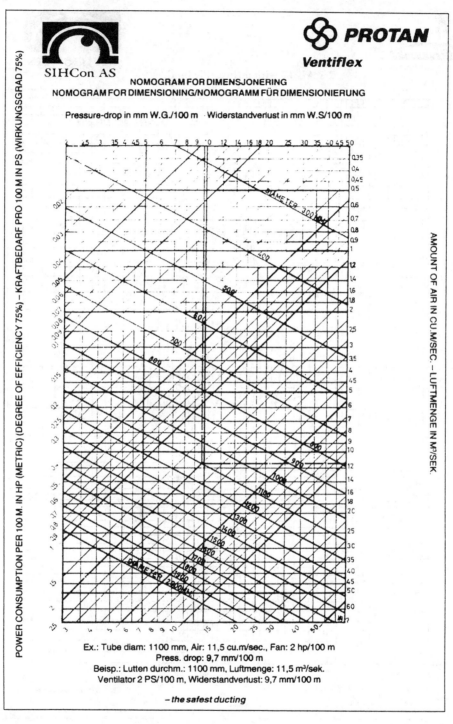

Figure 7. Monogram for calculations.

4. REVERSIBLE FLOW VENTILATION.

D& B Pressure drop 0 - 9100

Qin Ling   Calculated pressure drop in Ventiflex duct   D&B tunnel

Length [m]	Flowrate [m3/s]	Diameter, duct [m]	Area [m2]	Velocity [m/s]	Pressure drop, duct [Pa]	Power per 500 m
0 - 500	13,00	1,40	1,54	8,44	105,47	2,06
500 - 1000	12,48	1,40	1,54	8,11	97,77	1,83
1000 - 1500	11,96	1,40	1,54	7,77	90,33	1,62
1500 - 2000	11,43	1,40	1,54	7,43	83,17	1,43
2000 - 2500	10,91	1,40	1,54	7,09	76,29	1,25
2500 - 3000	10,39	1,40	1,54	6,75	69,67	1,09
3000 - 3500	9,87	1,40	1,54	6,41	63,34	0,94
3500 - 4000	9,35	1,40	1,54	6,07	57,28	0,80
4000 - 4500	8,82	1,40	1,54	5,73	51,50	0,68
4500 - 5000	8,30	1,40	1,54	5,39	46,01	0,57
5000 - 5500	7,78	1,40	1,54	5,05	40,80	0,48
5500 - 6000	7,26	1,40	1,54	4,71	35,88	0,39
6000 - 6500	6,74	1,40	1,54	4,38	31,25	0,32
6500 - 7000	6,21	1,40	1,54	4,04	26,92	0,25
7000 - 7500	5,69	1,40	1,54	3,70	22,89	0,20
7500 - 8000	5,17	1,40	1,54	3,36	19,15	0,15
8000 - 8500	4,65	1,40	1,54	3,02	15,73	0,11
8500 - 9000	4,13	1,40	1,54	2,68	12,62	0,08
9000 - 9100	3,60	1,40	1,54	2,34	1,97	0,01

Friction loss Pf=   948,02

Total Pressure Pt=   1137,62

Calculations are based on the formulae:

where: v= velocity in duct
d= duct diameter
l= length

$$\Delta p = 0.011 \cdot \frac{v^{1.85}}{d^{1.24}} \cdot l$$

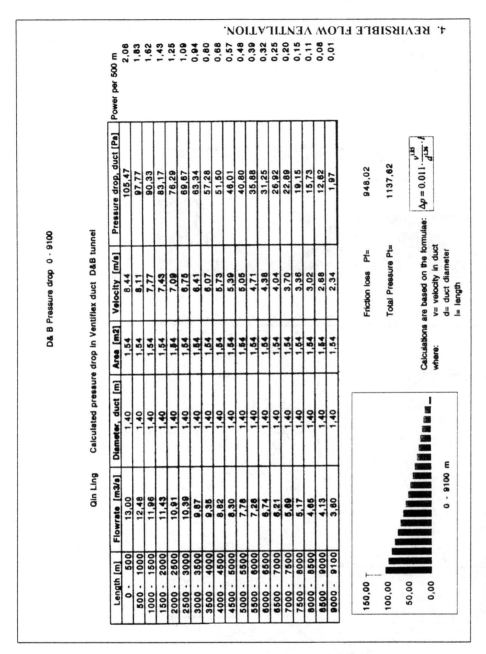

0 - 9100 m

Figure 8. Calculated pressure drop.

After thorough calculation and careful consideration the system was designed with two main fan-stations and 4 ductlines down the access tunnel, continuing with two ducts towards each tunnel front (see Fig. 13).

All the fans were remote controlled by wireless data modems.

Because reversible flow was essential in this system, the control had to be reliable and simple.

Figure 9.

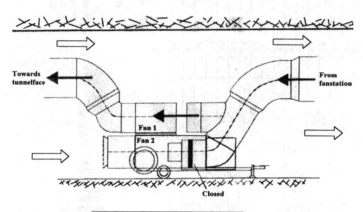

Figure 10.

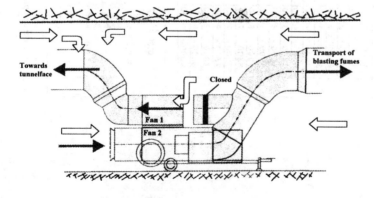

Figure 11.

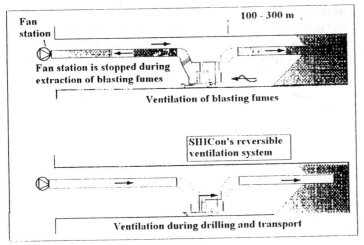

Figure 12.

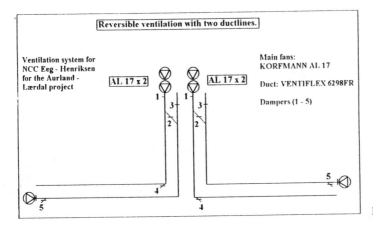

Figure 13.

Today, the length of the tunnel is about 7 km, and before it meets the tunnel from the Aurland side it will be about 9 km.

At each tunnel front there is a control panel, and by pressing different buttons for drilling, blasting, loading, transport and installation of ducts and a few other options. The signal is transmitted to the computer at the site office, which operates the total fan-system and dampers.

In the program the airflow is precalculated for all the operations and different lengths. The fans have three speed motors and seven different pitches of the blades.

The fansystem at the tunnelentrance had 920 kW installed power, and the cost for energy for the ventilation system alone would be high if no control was applied.

By using a flexible system with wireless control that monitors toxic gases the results should be optimum. In addition two ductlines reduce the pressure drop and leakage and eliminate booster fans.

This system has proven that investing in planning and control systems combined with computer technology has been rewarding.

We are at the moment developing the system further, based on what we have experienced in 'The world's longest road tunnel' and within a short period of time new products and systems will be presented which will improve today's tunnelventilation.

The level of intelligence of this system can always be improved, but the applied technology has to be tested and approved by the users.

Bulk emulsion explosives in tunnelling

JOHN JOHANSEN & C.F. MATHIESEN

In the 'good' old days dynamite was the only conceivable explosives in hard rock tunnelling – high weight strength, low critical diameter and water resistance. Later, the use of ANFO with a dynamite stick at the bottom gradually took over as a more cost efficient alternative. Lately, the bulk emulsion explosives have come into focus and are rapidly gaining ground.

Emulsion explosives in bulk form was for the first time tested in Norway in 1994/1995 in three different stages in a tunnel project on the west coast on the initiative of Dyno Nobel. The experience gained in these tests were so encouraging that The Norwegian Public Roads Administration decided that a complete tunnel was to be driven using bulk emulsion only. The Hanekleiv tunnel in the south-eastern part of Norway was chosen for this full scale test.

- Tunnel length   1730 M
- Cross-section (twin tunnel)   65 m^2 each
- No of unloaded holes in the cut (102 mm)   4
- No of loaded holes (45 mm), alt 1   87
- No of loaded holes (51 mm), alt 2   82
- Borehole depth, average   5.02 m
- Explosives consumption, alt 1   1.60 kg/m^3
- Explosives consumption, alt 2   1.83 kg/m^3
- Initiation   Nonel LP

One-way ventilation with AMV – fans, corresponding to GAL 14. 1100/1100 and canvas diameter 1500 mm.

The test was most successful. The following summary is taken from the test report which speaks for itself.

The Hanekleiv Tunnel in Vestfold, Norway is the first tunnel where emulsion explosives are used in regular tunnelling. Extensive follow-up work was done during the excavation of the tunnel. Emphasis was put on comparing the use of ANFO and emulsion explosives, with regard to blasting results, occupational environment and mechanised charging.

The measurements proved that the emulsion explosives and ANFO are equal in blasting performance. Emulsion explosives result in a considerably better occupational environment especially for the toxic fumes NO_2 and CO, and also for dust and visability. On

147

the basis of the good results, one may expect that the emulsion explosives will be the choice of future tunnelling.

Advantages and disavantages with bulk emulsions can best be summarized as follows:

Advantages
– The use of bulk emulsion results in better environmental conditions for the tunnel workers as compared with ANFO, especially with respect to toxic fumes, dust and visability.
– The blasting characteristics of emulsions as compared with ANFO vary little.
Emulsions are water resistant and well suited for declining tunnels and for tunnels with water problems.
– With the use of emulsions, only one type of explosives is in use.
– The bulk emulsion explosive is not a blasting agent prior to loading. Consequently no transport or storage of explosives takes place which means safety in handling.
– With emulsions, the amount of explosives in each borehole can be carefully controlled.
– New rules and regulations which prohibit simultaneous drilling and loading will give equal loading time for ANFO as for the emulsions.

Disadvantages
– Emulsions are more costly to procure.
– Regulations require that the explosives supplier has a responsible person present on site. This is an adding cost element provided the work force can not be reduced accordingly.
– Problems have been encountered with the emulsion being forced out of the borcholes by the water pressure.
Never-the-less, and in conclusion it is fair to assume that the bulk emulsions will be the dominating explosive in hard rock tunnelling in the years to come.

REFERENCE

The Use of Slurry Explosives in the Hanekleiv Tunnel in Vestfold. J. Elvøy, I Storås, P-E. Rønn.

Glossary of blasting terms

abutment – the point in a tunnel section where wall and roof meet
abutment height – height from tunnel floor to abutment
acceleration – unit of ground vibration in g (1 g = 9,81 m/sec2)
access tunnel – tunnel from surface to underground work site
adit – horizontal entrance to mine
advance – excavated length of tunnel per blasting round
airblast – airborne shockwave resulting from the detonation of explosives
ammonium nitrate (AN) – the most commonly used oxidizer in explosives and blasting
    agents
amplitude – unit of ground vibration (height of deflection in mm)
ampere – a unit of electrical current produced by 1 volt acting through a resistance of 1
    ohm
ANFO – an explosive material in powder form consisting of ammonium nitrate and fuel
    oil
arch height - height from abutment to the highest point of the tunnel roof

back break – the breakage that occurs beyond the last row of blast holes
base charge – main explosive charge in a detonator
blast – the act of fragmenting rock by the use of explosives
blast area – area close to a blast, which may be influenced by flyrock, concussion or air-
    borne shockwave
blaster – qualified person in charge and firing of a blast (see shot firer)
blasting machine – machine expressly built for initiating electric detonators (see exploder)
blasting vibration – the energy from a blast that manifests itself in earthborne vibrations
    that are transmitted through the earth away from the immediate blast area
booster – charge of high explosive used to improve detonations stability and to intensify
    the explosive reaction
bubble energy – energy of expanding gases of an explosive, as measured in an underwater
    test
bulk explosive – explosive material prepared for use without packing
bull hole – large empty center hole in a parallel hole cut
burden – distance from an explosive charge in a blasthole to the nearest free or open face
burn cut – type of parallel hole cut for tunnel blasting

careful blasting – blasting with respect to the surrounding rock

149

cautious blasting – blasting with respect to surrounding areas, controlling flyrock, ground vibrations and air shock waves

collar – opening of a blasthole

collar distance – distance from top of explosive to collar of blasthole

connecting wire – wire used to extend the firing line or legwires in an electric blasting circuit

contour holes – holes drilled along the perimeter of the excavation

controlled blasting – technique to control overbreak and damage to remaining rock surface

critical diameter – minimum diameter of an explosive for propagation of stable detonation

current leakage – portion of the firing current by-passing part of the blasting circuit through unintended paths

cut – opening part of a tunnel blast to provide a free face for the remainder of the round

cut easer holes – in tunnelling, the holes closest to the cut used to enlarge the opening formed by the cut

decibel – a unit of air overpressure commonly used to measure air blast

delay blasting – use of delay detonators to cause separate charges to detonate at different times

delay detonator – detonator, electric or non-electric, with a built-in delay element creating a delay between the input of energy and the explosion of the detonator

delay time – time between initiation and detonation of the base charge of a delay detonator

density – the mass of an explosive per unit of volume, usually expressed in grams per cubic centimeter (gr/c.c.)

detonating cord – a flexible cord containing a center core of high velocity explosive, used to initiate explosive charges

detonation – an explosive reaction that moves through an explosive material at a velocity greater than the speed of sound in the material and creates a high pressure shock wave, heat and gases

detonation pressure – the pressure produced in the reaction zone of a detonating explosive

detonation velocity – the velocity at which a detonation progress through an explosive

displacement – unit of ground vibration – height of deflection in mm

drilling pattern – plan of holes laid out on a tunnel face which are to be drilled for blasting. The burden and spacing are usually expressed in meters while the diameter of the blastholes is expressed in millimeters and inches

dynamite – high explosive invented by Alfred Nobel. Any high explosive containing nitroglycerin as a sensitizer is considered a dynamite

earth fault – see current leakage

electric detonator – detonator designed to be initiated by an electric current

emulsion – explosive where the oxidizer are dissolved in water and surrounded by immiscible fuels

exploder – see blasting machine

explosion – a chemical reaction involving an extremely rapid expansion of gases, usually associated with the liberation of heat

explosive – chemical mixture that releases gases and heat at high velocity, causing very high pressures

face – rock surface against which a blast can be executed

fan cut – cut for tunnel blasting where the opening holes are spread in the form of a fan

fault – natural crack formation in the rock

firing cable – cable connecting the blasting round with the blasting machine

flyrock – rocks propelled from the blast area by the force of an explosion

fragmentation – the breaking of a solid mass into pieces by blasting

frequency – unit of ground vibration characteristics – periods per second

fuel oil – fuel, usually diesel fuel – in ANFO

galvanometer – see ohm meter

ground vibration – shock wave emanating from a blast transmitted through the surrounding ground

half-second detonator – delay detonator with approximately 0,5 sec. delay between subsequent numbers

hertz(Hz) – term used to express the frequency of ground vibrations

instantaneous detonator – Detonator not containing any delay element

interval – difference in delay time between detonators with different numbers

joints – planes within the rock mass which separate solid rock mass from each other

leads – see leg wires

lead wire – see firing cable

leakage resistance – the resistance between the blasting circuit and the ground

leg wires – the two single wires or one duplex wire extending out from an electric detonator

lifters – blastholes in a tunnel round breaking upwards

loading density – the weight of explosive loaded per unit length of borehole occupied by the explosive, expressed as kilograms per meter of borehole

look-out – angling of the contour holes in a tunnel outside the theoretical contour to provide space for the drilling equipment when drilling the following round

millisecond detonator – short delay detonator with less than 100 ms delay between subsequent numbers

misfire – charge, or part of charge, which has failed to fire as planned

nitroglycerin – explosive oil originally used as sensitizer in dynamites

nonelectric detonator – a detonator that does not require the use of electric energy or safety fuse to function

Ohm meter – used to check the resistance of a single electric detonator, detonators in series and parallel and to check the final round. Has to be approved by the authorities for use in blasting operations

overburden – loose deposits laying on the top of rock

parallel hole cut – tunnel cut with all holes parallel and perpendicular to the rock face. The uncharged holes are normally larger than the blastholes

particle velocity – measure of ground vibration. The velocity at which a particle of ground vibrates when hit by a seismic wave

powder factor – specific charge

presplitting – preshearing – blasting of closely spaced holes along the perimeter of the excavation. The presplit is fired before the main blast

primer – cap-sensitive cartridge of high explosive which is used to initiate blasting agents

raise – drift which is vertical or has an angle of at least 45 upwards

relievers – cut easer holes

resistance – the measure of opposition to the flow of electrical current, expressed in ohm

round – group or set of blastholes forming a blast, when connected to each other

safety fuse – core of black powder covered by textile and water-proofing material, which is used to initiate fuse detonator

sensitizer – ingredient used in an explosive to ease initiation and propagation of detonation

short delay blasting – the practice of detonating blastholes in successive intervals where the time difference between any two successive detonations is measured in milliseconds

shot firer – person who actually fires the blast. This person is assigned to control the blasting operation with authority to decide charge weights, delay patterns etc.

sink shaft – underground shaft excavated vertically downwards

slurry explosive – high density aqueous explosive containing ammonium nitrate, sensitized with a fuel, thickened and cross-linked to a gelatinous consistency. Also called watergel

smoothblasting – method of controlled blasting in which closely spaced holes are drilled at the perimeter of the excavation, charged with low charges to reduce overbreak. The perimeter holes are fired with a higher delay number than the rest of the round

spacing – distance between blastholes in a row

specific charge – explosives consumption per cubic meter of rock

specific drilling – drilled meters per cubic meter of rock

tamping – act of compressing the explosive charge in a blasthole

tamping rod – rod of wood used to introduce and tamp explosive charge in a blasthole

V-cut – tunnel cut with the holes in V-layout. Also called wedge cut

vibration velocity – unit of ground vibration in mm/sec

volt – the unit of electromotive force. It is the difference in potential required to make a current of I ampere flow through a resistance of I ohm

Watt – a unit of electrical power equal to 1 joule/sec.

References

Atlas Powder Company 1987. Explosives and rock blasting. Technical Operations. Dallas, Texas. USA.

Berdal, B., Buen, B. & J. Johansen 1985. Lake Tap – The Norwegian Method, Paper 9, 6 pages. Tunnelling 85, Brighton, England. The Institution of Mining and Metallurgi 1985.

Cole, Robert H. 1965. *Underwater Explosions*. Dover Publications Inc. New York.

Dowding, Charles H. 1985. *Blast Vibration Monitoring and Control*. Prentice-Hall Inc., Englewood Cliffs, NJ.

Du Pont 1980. *Blasters Handbook*. 16th edition. E.I. du Pont de Nemours & Co. (Inc.) Explosives Products Division Wilmington, Delaware.

Dyno Industrier AS 1981. Sprengstoffer – Sprengningsteknikk. Dyno, Oslo

Gustafsson, Rune 1973. Swedish Blasting Technique. SPI, Gothenburg, Sweden

Hemphill, Gary B. 1981. *Blasting Operations*. McGraw-Hill Book Company, New York.

Holmberg, Roger 1982. *Vibrations generated by trafic and building construction activities*. Stockholm.

ICI Nobels Explosives company 1972. *Blasting Practice*. Fourth edition. ICI, Stevenston, Ayrshire, Scotland.

Langefors, U. & B. Kihlstrøm 1978. *The Modern Technique of Rock Blasting*. Third edition. Almqvist & Wiksell, Stockholm.

Meyer, Rudolf 1987. *Explosives*. Third edition. Weinheim, Germany.

Norwegian Soil and Rock Engineering Association 1981. Publication no. 1: *Norwegian Hard Rock Tunneling*. Tapir Publishers, University of Trondheim. Trondheim.

Norwegian Soil and Rock Engineering Association 1983. Publication no. 2: *Norwegian Tunnelling Technology*. Tapir Publishers.

Norwegian Soil and Rock Engineering Association 1985. Publication no. 3: *Norwegian Hydropower Tunnelling*. Tapir Publishers.

Norwegian Soil and Rock Engineering Association 1986. Publication no. 4: *Norwegian Road Tunnelling*. Tapir Publishers.

Norwegian Soil and Rock Engineering Association 1988. Publication no. 5: *Norwegian Tunnelling Today*. Tapir Publishers.

Norwegian Soil and Rock Engineering Association 1992. Publication no. 8: *Norwegian Subsea Tunnelling*. Rotterdam: Balkema.

Olofsson, Stig O. 1990. *Applied Explosives Technology for Construction and Mining*. Second edition. Applex, Sweden.

Solvik, Øivind 1981. Underwater piercing of a tunnel. Norwegian Hydrotechnical Laboratory, Trondheim.

University of Trondheim 1988. Publication 2-88: Tunnelling – Prognosis for Drill and Blast. NTH.

University of Trondheim 1988: Publication 3-88: Tunnelling – Costs for Drill and Blast. NTH.

University of Trondheim 1988. Publication 5-88: Rock Caverns. NTH.

University of Trondheim 1990. Publication 13-90: Drillability – Drilling Rate Index. Catalogue. NTH.

University of Trondheim 1994. Publication 1-94: Hard Rock Tunnel Boring. NTH.

University of Trondheim 1995. Publication 2A-95 Tunnelling – Blast Design. NTH.